D0205802

Fundamental Groups and Covering Spaces

Fundamental Groups and Covering Spaces

Elon Lages Lima

Translated by Jonas Gomes

A K Peters
Natick, Massachusetts

Editorial, Sales, and Customer Service Office

A K Peters, Ltd.
63 South Avenue
Natick, MA 01760
www.akpeters.com

Library of Congress Cataloging-in-Publication Data

Lima, Elon Lages, 1929-
[Grupo fundamental e espacos de recobrimento. English]
Fundamental groups and covering spaces / Elon Lages Lima ; translated by Jonas Gomes.
p. cm.
Includes bibliographical references and index.
ISBN 1-56881-131-4
1. Algebraic topology. 2. Fundamental groups (Mathematics) 3. Covering Spaces (Topology) I. Title.

QA612.L4713 2003
514'.2--dc21

2003048719

Printed in Canada
07 06 05 04 03 10 9 8 7 6 5 4 3 2 1

Contents

Preface

The origin of this book is a set of notes for a course that I taught, years ago, in the Brazilian Mathematical Colloquium and, several times after that, to beginning graduate students at IMPA, Rio de Janeiro. Later on, the notes were revised and appeared as a book in the collection "Projeto Euclides," published by IMPA. Since then the book has been used as an introduction to algebraic topology in many Brazilian universities and in other Latin American countries.

The subjects discussed here in, fundamental group and covering spaces, are well suited as an introduction to algebraic topology for their elementary character, for exhibiting in a clear way the use of algebraic invariants in topological problems and also because of the immediate applications to other areas of mathematics such as real analysis, complex variables, differential geometry and so on.

This is an introductory book, with no claims of becoming a reference work. The appeals to facts of analysis and algebra that are made in the text are very few and their aim is to show connections with other disciplines. If the reader so wishes, these appeals may be skipped without harm to the understanding of the text.

It is a pleasure to extend my warmest thanks to Jonas Gomes, a very dear friend and colleague, who suggested the translation of the book into English and, to my great surprise and contentment, undertook the job himself with his habitual competence, recommending a few changes and additions, which I made with satisfaction.

Rio de Janeiro, January 2003
Elon Lages Lima

Part I
Fundamental Groups

"*Quant à moi, toutes les voies diverses où je m'étais engagé successivement me conduisaient à l'Analysis Situs. J'avais besoin des donnés de cette Science pour poursuivre mes études sur les courbes définies par les équations différentielles et pour les étendre aux équations différentielles d'ordre superieur et, en particulier, à celles du problème de trois corps. J'en avais besoin pour l'étude des fonctions non uniformes de deux variables. J'en avais besoin pour l'étude des périodes des intégrales multiples et pour l'application de cette étude au developpement de la fonction perturbatrice. Enfin, j'entrevoyais dans l'Analysis Situs un moyen d'aborder un problème important de la théorie des groupes, la recherche des groupes discrets ou des groupes finis contenus dans un groupe continu donné.*"

H. Poincaré (Acta Mathematica, vol. 38 (1921) pp. 101.)

Henri Poincaré (1854–1912), the extraordinary French mathematician, was considered "the last universalist," that is, a contributor to the progress of all important areas of mathematics. We owe to him the notion of fundamental groups and the creation of homology theory, which are fundamental concepts of topology.

The above quotation, from his scientific autobiography ("Notice sur les travaux scientifiques de Henri Poincaré"), was published for the first time nine years after his death, in a special issue of the journal "Acta Mathematica," dedicated to him.

Chapter 1
Homotopy

Note: throughout this book, the symbol I will denote the compact interval $[0,1]$ of real numbers.

In this chapter, we introduce the basic notions about homotopy that will be used throughout the book. Homotopy is, indeed, the most important idea of algebraic topology and the fundamental group—which we study in this book—is probably the simplest algebraic invariant associated to this idea. The fundamental group will be presented in the next chapter. We cover in this chapter general results related with homotopy, illustrating these concepts with applications and elementary examples. In particular, we show the connection between homotopy and the problem of extending a continuous map defined on a closed subset of a topological space.

1.1 Homotopic Maps

Let X, Y be two topological spaces. Two continuous maps $f, g\colon X \to Y$ are said to be *homotopic* when there exists a continuous map

$$H\colon X \times I \to Y$$

such that $H(x,0) = f(x)$ and $H(x,1) = g(x)$ for all $x \in X$. The map H is called a *homotopy* between f and g. We use the notation $H\colon f \simeq g$, or simply $f \simeq g$.

For each $t \in I$, the homotopy $H\colon f \simeq g$ defines the continuous map $H_t\colon X \to Y$, with $H_t(x) = H(x,t)$. This means that defining a homotopy H turns out to be equivalent to prescribing a continuous one-parameter family $(H_t)_{t\in I}$ of maps from X to Y. We have $H_0 = f$ and $H_1 = g$; therefore, the family $(H_t)_{t\in I}$ starts with f and ends at g.

Intuitively, the parameter t can be interpreted as being the time. The homotopy is considered a continuous deformation process of the map f. This deformation occurs during the unit of time. In the instant $t = 0$, we have f, and for $t = 1$, we get g. In the intermediate times, $0 < t < 1$, the maps H_t provide the intermediate stages of the deformation.

Example 1.1. Two constant maps $f, g\colon X \to Y, f(x) = p, g(x) = q$, are homotopic if, and only if, p and q belong to the same pathwise connected component of the space Y. Indeed, if there exists a path $a\colon I \to Y$ with $a(0) = p$ and $a(1) = q$, we define a homotopy $H\colon X \times I \to Y$ between f and g by $H(x,t) = a(t)$, for all $(x,t) \in X \times I$. Conversely, if H is a homotopy between the constant maps $f(x) = p$ and $g(x) = q$, by fixing arbitrarily $x_0 \in X$, we define a path $a\colon I \to Y$ connecting p to q by setting $a(t) = H(x_0, t)$. ◁

Example 1.2. Let $Y \subset E$, where E is a normed vector space. Given the continuous maps $f, g\colon X \to Y$, suppose that for all $x \in X$, the line segment $[f(x), g(x)]$ is contained in Y. Then $f \simeq g$. Indeed, we just have to define $H(x,t) = (1-t)f(x) + tg(x)$ to obtain a homotopy $H\colon X \times I \to Y$ between f and g. This is called a *linear homotopy*. For each $x \in X$ fixed and t varying from 0 to 1, the point $H(x,t)$ moves (with uniform velocity) on the line segment connecting $f(x)$ to $g(x)$. As particular cases, we obtain the statements A and B below.

A. *Any two continuous maps $f, g\colon X \to E$ which take values on a normed vector space E are homotopic.*

In particular, every continuous map $f\colon X \to E$ is homotopic to the constant map 0, by the homotopy $H(x,t) = (1-t)f(x)$.

B. (Poincaré-Bohl) *If $f, g\colon X \to E - \{0\}$ are two maps satisfying $|f(x) - g(x)| < |f(x)|$ for all $x \in X$, then $f \simeq g$.*

Indeed, if 0 belonged to the segment $[f(x), g(x)]$ for some $x \in X$, we would have

$$|f(x) - g(x)| = |f(x)| + |g(x)| \geq |f(x)|.$$

Therefore, $[f(x), g(x)] \subset E - \{0\}$ for all $x \in X$; hence, f is linearly homotopic to g. ◁

From Chapter 3 on, we will be able to exhibit interesting examples of non homotopic maps. At present, we will be contented with this remark: If $f, g\colon X \to Y$ are continuous maps whose images $f(X)$ and $g(X)$ are contained in distinct connected components of Y, then f and g are not

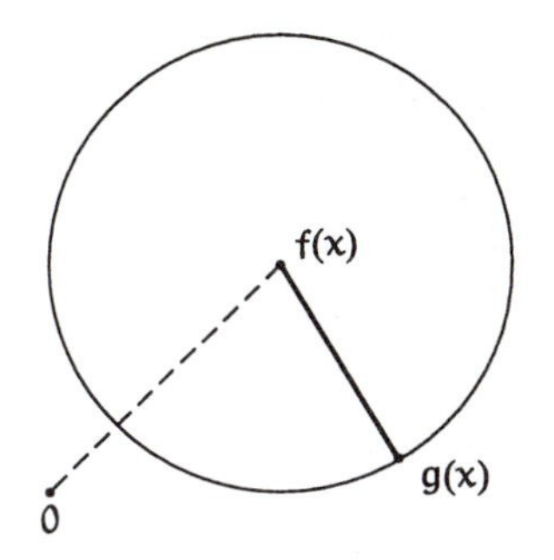

Figure 1.1. When $|f(x) - g(x)| < |f(x)|$, the line segment $[f(x), g(x)]$ does not contain the origin 0.

homotopic. This example is not very interesting because, in the theory of homotopy, it is usual to consider mostly pathwise connected topological spaces, as we will see in the chapters to follow.

The range Y plays an important role in the homotopies. It is the space where the deformation takes place. By increasing Y, we can allow new homotopies. If $Y \subset Y'$, it may occur that two continuous maps $f, g\colon X \to Y$ are not homotopic but, considered as maps from X to Y', they are. For example, any two maps $f, g\colon X \to E$, taking values in a normed vector space, are always homotopic but the same does not occur for all $f, g\colon X \to E - \{0\}$. We just have to take $f, g\colon \mathbb{R} \to \mathbb{R} - \{0\}$ constant maps, with $f(x) = 1$ and $g(x) = -1$ for all x. Since 1 and -1 belong to distinct connect components of $\mathbb{R} - \{0\}$, it follows that f and g are not homotopic.

Proposition 1.1. *Consider two topological spaces X and Y. The homotopy relation $f \simeq g$ is an equivalence relation in the set of continuous maps from X to Y.*

Proof. For every continuous map $f\colon X \to Y$, the map $H\colon X \times I \to Y$, defined by $H(x,t) = f(x)$, is a homotopy between f and f, hence $\simeq$ is reflexive. Now consider the homotopy $H\colon X \times I \to Y$ between f and g. By defining $K\colon X \times I \to Y$ by $K(x,t) = H(x, 1-t)$, it is easy to verify that K is a homotopy between g and f. Hence, $f \simeq g \Rightarrow g \simeq f$; that is, the homotopy relation is symmetric. Finally, if $H\colon f \simeq g$ and $K\colon g \simeq h$, then we define $L\colon X \times I \to Y$ by $L(x,t) = H(x, 2t)$ if $0 \leq t \leq 1/2$ and $L(x,t) = K(x, 2t-1)$ if $1/2 \leq t \leq 1$. The map L is a homotopy between f and h. Hence, $f \simeq g, g \simeq h \Rightarrow f \simeq h$; that is, the relation $\simeq$ is transitive. □

The equivalence classes of the homotopy relation are called *homotopy classes.* The homotopy class of a continuous map $f\colon X \to Y$ is represented by the symbol $[f]$. The set of the homotopy classes of the continuous maps from X to Y is represented by the symbol $[X, Y]$.

Remark. Consider the topological space $C(X;Y)$, of all continuous maps from X to Y, with the compact-open topology. (When Y is metrizable, this is the topology of uniform convergence on the compact subsets of X.) To each map $H\colon X \times I \to Y$ there corresponds a path in $C(X;Y)$; that is, a map $\widetilde{H}\colon I \to C(X;Y)$ defined by $\widetilde{H}(t) = H_t, H_t(x) = H(t,x)$. When X is Hausdorff locally compact or metrizable then H is continuous if, and only if, $\widetilde{H}$ is. Hence, for X metrizable, or Hausdorff locally compact, there exists a natural bijection between the homotopies $H\colon X \times I \to Y$ and the paths $\widetilde{H}\colon I \to C(X;Y)$. If H is a homotopy between f and g then the path $\widetilde{H}$ starts in $f = H_0$ and ends in $g = H_1$. It follows that two continuous maps $f, g\colon X \to Y$, with X locally compact Hausdorff or metrizable, are homotopic if, and only if, f and g belong to the same pathwise connected component of the space $C(X;Y)$. Therefore, for such spaces X, the homotopy classes of maps $f\colon X \to Y$ are the pathwise connected components of the space $C(X, Y)$. For additional information about the compact-open topology, see Bredon (1993), page 437.

Proposition 1.2. *Let $f, f'\colon X \to Y$ and $g, g'\colon Y \to Z$ be continuous maps. If $f \simeq f'$ and $g \simeq g'$, then $g \circ f \simeq g' \circ f'$. In other words: map composition preserves homotopies.*

Proof. Let $H\colon X \times I \to Y$ be a homotopy between f and f', and $K\colon Y \times I \to Z$ a homotopy between g and g'. We define a homotopy $L\colon X \times I \to Z$, between $g \circ f$ and $g' \circ f'$, by $L(x,t) = K(H(x,t),t)$. □

By Proposition 1.2, we can define the operation of composition between homotopy classes. Given $f\colon X \to Y$ and $g\colon Y \to Z$, we define $[g] \circ [f] = [g \circ f]$. The class $[g \circ f]$ does not depend on the representatives g, f of the classes $[g]$ and $[f]$, respectively.

1.1.1 Vector Fields on Spheres

In this section, we will use the concept of homotopy to investigate the existence of non-null tangent vector fields on the unit sphere $S^n \subset \mathbb{R}^{n+1}$.

Proposition 1.3. *Given two continuous maps $f, g\colon X \to S^n$, if $f(x) \neq -g(x)$ for all $x \in X$ (that is, if $f(x)$ and $g(x)$ are never antipodal points), then $f \simeq g$.*

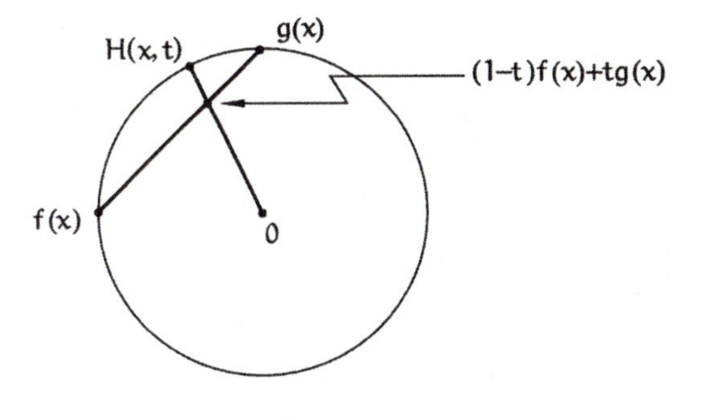

Figure 1.2.

Proof. Indeed, under these conditions we have $(1-t)f(x)+tg(x) \neq 0$ for all $t \in I$ and all $x \in X$. Hence, we obtain a homotopy $H\colon X \times I \to S^n$, between f and g, by taking the radial projection

$$H(x,t) = \frac{(1-t)f(x)+tg(x)}{|(1-t)f(x)+tg(x)|}.$$

When t varies between 0 and 1, $H(x,t)$ describes the shortest arc of the great circle that connects $f(x)$ to $g(x)$ (see Figure 1.2). □

The condition that $f(x)$ and $g(x)$ are never antipodal can be expressed by $|f(x)-g(x)| < 2$ for all $x \in X$.

As particular cases of the proposition, we obtain:

A. If $f\colon S^n \to S^n$ has no fixed points (that is, $f(x) \neq x$ for all x), then f is homotopic to the antipodal map $\alpha\colon S^n \to S^n, \alpha(x) = -x$.

B. If $f\colon S^n \to S^n$ satisfies $f(x) \neq -x$ for all $x \in S^n$, then f is homotopic to the identity map of S^n.

Proposition 1.4. *If n is odd, then the antipodal map $\alpha\colon S^n \to S^n$, $\alpha(x) = -x$ is homotopic to the identity id: $S^n \to S^n$.*

Proof. Let $n = 2k-1$. Then $S^n \subset \mathbb{R}^{2k}$ and we may consider each point

$$z = (x_1, y_1, x_2, y_2, \ldots, x_k, y_k)$$

in S^n as a list $z = (z_1, \ldots, z_k)$ of complex numbers $z_j = x_j + i \cdot y_j$ such that

$$|z_1|^2 + \cdots + |z_k|^2 = 1.$$

For each complex number $u \in S^1$, of modulus 1, and each vector $z = (z_1, \ldots, z_k) \in S^n$ we define $u \cdot z \in S^n$ by $u \cdot z = (u \cdot z_1, \ldots, u \cdot z_k)$. With this notation, $H\colon S^n \times I \to S^n$, defined by

$$H(z,t) = e^{t\pi i} \cdot z,$$

is a homotopy between the antipodal map $\alpha(z) = -z$ and the identity map of S^n. □

The converse of Proposition 1.4 is true, as stated next

Proposition 1.5. *If n is even, the antipodal map $\alpha\colon S^n \to S^n$, $\alpha(x) = -x$ is not homotopic to the identity map of S^n.*

Proof. We will only sketch the proof. It does not use any of the previous results about homotopy, but is based on some analysis concepts.

When $n = 2k$ is even, we have $S^n \subset \mathbb{R}^{2k+1}$ and the antipodal map

$$\alpha(x_1, \ldots, x_{2k+1}) = (-x_1, \ldots, -x_{2k+1})$$

has determinant equal to -1, so it reverses orientation (the concept of orientation will be studied in detail in Chapter 8 of this book). If we had a homotopy $\alpha \simeq Id\colon S^n \to S^n$, then, considering the volume element dV in S^n, we would have

$$\int_{S^n} \alpha^* dV = \int_{S^n} (Id)^* dV = \int_{S^n} dV = \text{Volume of } S^n.$$

Since, however, α is orientation reversing, we have

$$\int_{S^n} \alpha^* dV = -\int_{S^n} dV = -(\text{Volume of } S^n).$$

For more details, see Bredon (1993), page 265. □

Proposition 1.6. *If there exists a non-null continuous vector field on S^n, then the antipodal map $\alpha\colon S^n \to S^n$ is homotopic to the identity.*

Proof. Given $v\colon S^n \to \mathbb{R}^{n+1}$ continuous, tangent, and non-null at every point, we define $f\colon S^n \to S^n$ by the radial projection

$$f(x) = \frac{x + v(x)}{|x + v(x)|}.$$

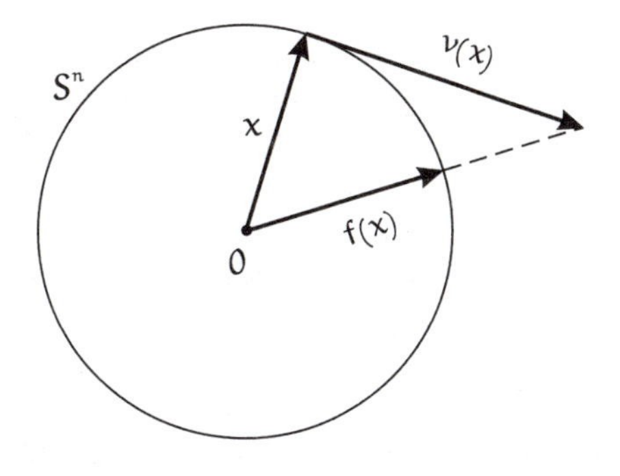

Figure 1.3.

Then f is continuous and $f(x) \neq x$ for all x. Hence, $f \simeq \alpha$. On the other hand, the map $H\colon S^n \times I \to S^n$, defined by

$$H(x,t) = \frac{x + t \cdot v(x)}{|x + t \cdot v(x)|},$$

is a homotopy H between the identity map of S^n and f. By transitivity it follows that $\alpha \simeq$ identity. $\square$

When $n = 2k - 1$ is odd, there exists a non-null continuous tangent vector field $v\colon S^n \to S^n$, defined by

$$v(x_1, \dots, x_k, y_1, \dots, y_k) = (-y_1, \dots, -y_k, x_1, \dots, x_k).$$

In this way, we reobtain the result in Proposition 1.4: n odd $\Rightarrow \alpha \simeq id\colon S^n \to S^n$.

The results proven in Propositions 1.3, 1.4, and 1.6 are indicated by $\Rightarrow$ on the diagram in Figure 1.4. The implication on the diagram indicated

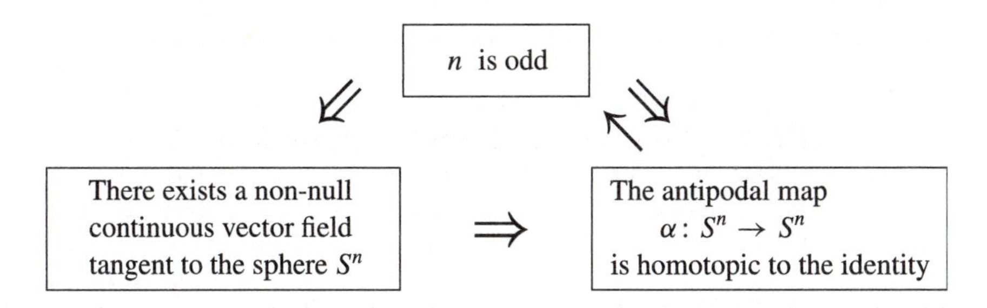

Figure 1.4. Vector fields and homotopies on the sphere.

by $\rightarrow$ is the statement of Proposition 1.5 (of which we have just sketched the proof here). If follows that all three statements on the diagram are equivalent, and we have the following proposition.

Proposition 1.7. *There exists a non-null continuous vector field tangent to the unit sphere S^n if, and only if, n is odd.*

1.2 Homotopy Type

A continuous map $f\colon X \rightarrow Y$ is called a *homotopy equivalence* when there exists a continuous map $g\colon Y \rightarrow X$ such that $g \circ f \simeq id_X$ and $f \circ g \simeq id_Y$. When this happens, we say that g is a *homotopy inverse* of f and that the topological spaces X and Y have the same *homotopy type*. In this case, we will write $X \equiv Y$ or $f\colon X \equiv Y$.

It is easy to see that the homotopy type defines an equivalence relation: $X \equiv X$, $X \equiv Y \Rightarrow Y \equiv X$ and also $X \equiv Y, Y \equiv Z \Rightarrow X \equiv Z$.

Example 1.3. The unit sphere S^n has the same homotopy type of $\mathbb{R}^{n+1} - \{0\}$. In fact, considering the inclusion map $i\colon S^n \rightarrow \mathbb{R}^{n+1} - \{0\}$, $i(x) = x$, and the radial projection $r\colon \mathbb{R}^{n+1} - \{0\} \rightarrow S^n$, $r(y) = y/|y|$, we have $r \circ i = id_{S^n}$. Moreover, $i \circ r\colon \mathbb{R}^{n+1} - \{0\} \rightarrow \mathbb{R}^{n+1} - \{0\}$ is homotopic to the identity map of $R^{n+1} - \{0\}$ using the linear homotopy, because every point $y \neq 0$ in $\mathbb{R}^{n+1}$ can be connected to $y/|y|$ by a line segment that does not contain the origin. $\triangleleft$

The same argument used in example above shows that if B^{n+1} is the closed ball of center 0 and radius 1 in $\mathbb{R}^{n+1}$ then $B^{n+1} - \{0\}$ has the same homotopy type of the sphere S^n.

If $h\colon X \rightarrow Y$ is a homeomorphism, it is evident that X and Y have the same homotopy type. The previous example shows that the converse is far from being true. In fact, homotopy type is a weaker topological invariant than homeomorphism. Intuitively, some points may get collapsed during the homotopy deformation that establishes the homotopy equivalence.

Example 1.4. Let $T = S^1 \times S^1$ be the bidimensional torus and $p \in T$ an arbitrary point. $T - \{p\}$ has the same homotopy type of the union of two circles with a point in common. Indeed, the torus T is the image of a square Q by a continuous map $\varphi\colon Q \rightarrow T$ that transforms the boundary ∂Q of the square in the union $Y = S \cup S'$ of two circles with a point in common, and it is a homeomorphism of the interior of Q onto the complement $T - Y$. We may assume that the point $p \in T$ is the image by φ of the center p_0 of the square. Since the square is homeomorphic to a closed unit disk, it

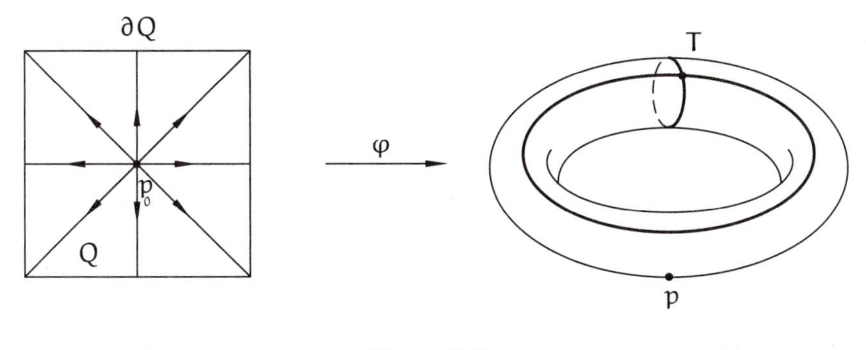

Figure 1.5.

follows that $Q - \{p_0\}$ has the same homotopy type of the boundary ∂Q. The deformation of $Q - \{p_0\}$ into ∂Q takes place along the rays that start at p_0. During the homotopy, the points of ∂Q do not move. By composing this homotopy with φ, we obtain a deformation of $T - \{p\}$ into Y, which gives us $T - \{p\} \equiv Y$. ◁

Example 1.5. If $X \equiv X'$ and $Y \equiv Y'$, then $[X, Y]$ and $[X', Y']$ possess the same cardinal number. More precisely, if $\varphi\colon X' \equiv X$ and $\psi\colon Y \equiv Y'$, then $[f] \mapsto [\psi \circ f \circ \varphi]$ is a bijection between $[X, Y]$ and $[X', Y']$. ◁

1.3 Contractible Spaces

A topological space X is called *contractible* when it has the same homotopy type of a point.

Proposition 1.8. *X is contractible if, and only if, the identity map $id\colon X \to X$ is homotopic to a constant map $X \to X$.*

Proof. If $f\colon X \to \{p\}$ is a homotopy equivalence and $g\colon \{p\} \to X$ is a homotopy inverse of f then $g \circ f \simeq id_X$ and $g \circ f$ is a constant map. Conversely, if $id_X \simeq$ constant, then id_X and the constant map are homotopy equivalences, one being inverse of the other. □

Corollary 1.1. *A contractible space X is pathwise connected.*

Indeed, if H is a homotopy between id_X and the constant map $X \to \{p\}, p \in X$, then, for each point $x \in X$, the correspondence $t \mapsto H(x, t)$ defines a path connecting x to p.

Proposition 1.9. *If either X or Y is contractible, then every continuous map $f\colon X \to Y$ is homotopic to a constant.*

Proof. If X is contractible and $H\colon X\times I \to X$ is a homotopy between id_X and a constant, then, for any $f\colon X \to Y$, the map $f\circ H\colon X\times I \to Y$ will be a homotopy between f and a constant. If Y is contractible and $K\colon Y\times I \to Y$ is a homotopy between id_Y and a constant, then $L\colon X \times I \to Y$, defined by $L(x,t) = K(f(x),t)$, is a homotopy between $f\colon X \to Y$ and a constant map. □

Corollary 1.2. *If X is contractible and Y is pathwise connected then any two continuous maps $f, g\colon X \to Y$ are homotopic. If Y is contractible then, for any space X, two continuous maps $f, g\colon X \to Y$ are always homotopic* (see Example 1.1).

Example 1.6. (Stars) A subset X of a normed vector space E is called a *star* with vertex p when, for all $x \in X$, the line segment $[p,x]$ is contained in X. If X is a star with vertex p then $H\colon X\times I \to X, H(x,t) = (1-t)x+tp$ is a homotopy between id_X and the constant map $X \to \{p\}$. Hence, every star is contractible. ◁

Example 1.7. (Convex sets) A subset $X \subset E$ of a normed vector space E is said to be *convex* when, for any two points $x, y \in X$, the line segment $[x,y]$ is contained in X. A convex set can be considered as a star with vertex at any of its points, and therefore, it is contractible. In particular, a normed vector space E is convex and therefore contractible. This explains (see Example 1.2) why any two continuous maps taking values on E are always homotopic. An open ball $B = B(a;r) \subset E$ is also convex. Indeed, if $x, y \in B$ and $0 \le t \le 1$ then $|(1-t)x + ty - a| = |(1-t)(x-a) + t(y-a)| \le (1-t)|x-a| + t|y-a| < (1-t)r + t\cdot r = r$; hence, $(1-t)x + ty \in B$. Therefore, $x, y \in B \Rightarrow [x,y] \subset B$. In the same way one shows that a closed ball $B[a;r]$ is also convex. ◁

Example 1.8. If the space X is contractible, then, for every Y, the Cartesian product $X\times Y$ has the same homotopy type as Y. In order to prove this fact, consider a homotopy H between id_X and the constant map $X \to \{p\}, p \in X$. Then the maps $f\colon X\times Y \to Y$, $f(x,y) = y$ and $g\colon Y \to X\times Y, g(y) = (p,y)$ are homotopy equivalences, because $f\circ g = id_Y$ and, moreover, $K(x,y,t) = (H(x,t),y)$ defines a homotopy between the identity map of $X\times Y$ and the map $g\circ f\colon X\times Y \to X\times Y$. In particular, if X and Y are both contractible, the Cartesian product $X\times Y$ is also contractible. ◁

1.4 Homotopy and Map Extension

One of the most important problems of topology is that of extending continuous mappings. Given a continuous map $f\colon A \to Y$, defined on a closed subset A of a topological space X, this problem consists of investigating the possibility of extending continuously f to X, that is, of the existence of a continuous map $\bar{f}\colon X \to Y$ such that $\bar{f}|A = f$.

The well-known Extension theorem of Tietze and Urysohn provides an affirmative answer to this problem when X is a normal space and Y is an interval of real numbers.

The extension problem is strongly related with the concept of homotopy, as we will show. In particular, we will show that the Tietze-Urysohn Theorem is still valid if, instead of a line interval, Y is a contractible space of the type ENR, to be defined later.

One of the simplest connections between homotopy and the extension of continuous maps is given by the proposition below, which, although elementary, is very useful.

Proposition 1.10. *A continuous map $f\colon S^n \to X$ extends continuously to the unit closed ball B^{n+1} if, and only if, it is homotopic to a constant.*

Proof. Consider the continuous map $\varphi\colon S^n \times I \to B^{n+1}$ defined by $\varphi(x,t) = (1-t)x$. We have $\varphi(x,0) = x$ and $\varphi(x,1) = 0$, for all $x \in S^n$.

Now suppose that $\bar{f}\colon B^{n+1} \to X$ is a continuous extension of $f\colon S^n \to X$. Consider the map $H = \bar{f} \circ \varphi\colon S^n \times I \to X$

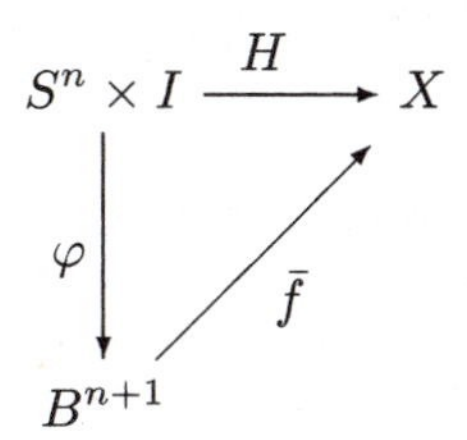

The map H provides a homotopy between f and the constant map $S^n \to f(0)$.

Conversely, suppose that $H\colon S^n \times I \to X$ is a homotopy between f and a constant map: $H(x,0) = f(x)$ and $H(x,1) = x_0 \in X$. Each point $y \in B^{n+1} - \{0\}$ can be uniquely written as $y = (1-t)x$, with $x \in S^n$, and $t \in [0,1)$. Define $\bar{f}\colon B^{n+1} \to X$ by

$$\bar{f}(y) = \begin{cases} \bar{f}((1-t)x) = H(x,t) & \text{if } y \neq 0 \\ x_0 & \text{if } y = 0. \end{cases}$$

For $x \in S^n$, we have $\bar{f}((1-0)x) = H(x,0) = f(x)$; thus, $\bar{f}$ is indeed an extension of f.

It remains to show that $\bar{f}$ is continuous. The reader can prove that if $K, L \subset \mathbb{R}^m$ are compact sets and $\varphi\colon K \to L$ is continuous and surjective, then a map $g\colon L \to X$ is continuous if, and only if, $g \circ \varphi\colon K \to X$ is continuous. Since $H = \bar{f} \circ \varphi$ is continuous, it follows that $\bar{f}$ is continuous, which completes the proof. □

It follows from the proof that if $H\colon S^n \times I \to X$ is a homotopy between f and a constant $p \in X$, the extension $\bar{f}\colon B^{n+1} \to X$ of f to the unit ball is given by

$$\bar{f}(y) = \begin{cases} H\left(\frac{x}{|x|}, 1 - |x|\right) & \text{if } x \in B^{n+1} - \{0\} \\ p & \text{if } x = 0. \end{cases}$$

Remarks. 1. The above proposition will be used in the following chapters with $n = 1$: A continuous map $f\colon S^1 \to X$ is homotopic to a constant if, and only if, it can be continuously extended to the closed unit disk of the plane.

2. When $n = 1$, we will use the proposition with a rectangle $I \times J$ instead of the disk B^2, and the boundary $\partial(I \times J)$ instead of the circle S^1.

1.4.1 Euclidean Neighborhood Retracts

We will now show that, under certain conditions, the possibility of continuously extending $f\colon A \to Y$ depends only on the homotopy class of f. For this, we will first introduce some concepts related with the notion of retract, due to K. Borsuk.

Let X be a topological space and Y a subspace of X. A continuous map $r\colon X \to Y$ is called a *retraction* when $r(y) = y$ for all $y \in Y$; that is, when $r|Y = id_Y$. Therefore, a retraction $r\colon X \to Y$ is a continuous extension to X of the identity map $Y \to Y$. Every retraction is surjective.

We have $Y = \{y \in X; r(y) = y\}$. It follows that, when X is a Hausdorff space, every retract $Y \subset X$ is a closed subset of X.

When there exists a retraction $r\colon X \to Y$, the subspace Y is called a *retract* of the space X.

Example 1.9. Every point is a retract of any space that contains it. For every $x_0 \in X$, the subspace $\{x_0\} \times Y$ is a retract of the product space $X \times Y$. If X is connected and Y is a disconnected subset of X, then Y is not a retract of X. If $X = X_1 \cup X_2$ is the union of two closed subsets with a single point in common, then X_1 and X_2 are both retracts of X. The radial projection $r\colon \mathbb{R}^{n+1} - \{0\} \to S^n$, $r(x) = x/|x|$, is a retraction. We will see in Chapter 3 that a circle in the plane is not a retract of the disk of which it is the boundary. It is true, more generally, that a sphere $S^n \subset \mathbb{R}^{n+1}$ is not a retract of the closed ball B^{n+1} which has the sphere as the boundary. This will be proved here for $n = 1$ (see Chapter 3, Section 2). For the general case, see Spanier (1966), page 194. ◁

A continuous map $r\colon X \to X$ such that $r \circ r = r$ is a retraction of X onto the subspace $Y = r(X)$, as can easily be verified.

Let $Y \subset \mathbb{R}^n$. We will say that Y is a *Euclidean neighborhood retract*, or that Y is of *type* ENR, when there exists a retraction $r\colon V \to Y$, where V is a neighborhood of Y in $\mathbb{R}^n$.

Example 1.10. If $Y \subset \mathbb{R}^n$ is a retract of $\mathbb{R}^n$, then Y is a retract of any of its neighborhoods in $\mathbb{R}^n$. On the other hand, the sphere $S^n \subset \mathbb{R}^{n+1}$ is not a retract of $\mathbb{R}^{n+1}$ but it is a retract of its neighborhood $V = \mathbb{R}^{n+1} - \{0\}$. Hence, S^n is of type ENR. ◁

Example 1.11. Every differentiable surface $M \subset \mathbb{R}^n$ is of type ENR. To see this, consider a tubular neighborhood V of M in $\mathbb{R}^n$. The projection $\pi\colon V \to M$ is a retraction. Every polyhedron $P \subset \mathbb{R}^n$ is also of type ENR. (See Eilenberg & Steenrod (1952), page 70.) These two classes are enough to indicate how numerous the spaces of type ENR are. ◁

Proposition 1.11. *Let $Y \subset \mathbb{R}^n$ be a compact space of type ENR. There exists $\varepsilon > 0$ such that any two continuous maps $f, g\colon X \to Y$, satisfying the condition $|f(x) - g(x)| < \varepsilon$ for every $x \in X$, are homotopic.*

Proof. Let $r\colon V \to Y$, a retraction of a neighborhood $V \supset Y$. Since Y is compact, we have $\varepsilon = \operatorname{dist}(Y, \mathbb{R}^n - V) > 0$. This means that if $y \in Y, z \in \mathbb{R}^n$ and $|y - z| < \varepsilon$, then $z \in V$. Under these conditions, the points of the line segment $[y, z]$ belong to V. Given $f, g\colon X \to Y$ with $|f(x) - g(x)| < \varepsilon$ for any $x \in X$, we have therefore $(1-t)f(x) + tg(x) \in V$ for every $x \in X$ and every $t \in I$. It follows that $H\colon X \times I \to Y$, defined by $H(x) = r[(1-t)f(x) + tg(x)]$, is a homotopy between f and g. □

Corollary 1.3. *Let M, N be differentiable compact surfaces. Every continuous map $f\colon M \to N$ is homotopic to a differentiable map.*

Indeed, take the $\varepsilon > 0$ that is attributed to N by Proposition 1.11. Applying the approximation theorem of continuous maps by differentiable ones, we obtain $g\colon M \to N$ differentiable, with $|f(x) - g(x)| < \varepsilon$ for every $x \in M$. Then f and g are homotopic.

Remarks. 1. By Proposition 1.3, for $Y = S^{n-1}$, the number ε of the above proposition is equal to 2.

2. The compactness hypothesis in Proposition 1.11 can be eliminated. In this case we obtain, instead of a constant ε, a continuous function $\varepsilon\colon Y \to \mathbb{R}^+$, such that if $f, g\colon X \to Y$ satisfies $|f(x) - g(x)| < \varepsilon(f(x))$ for every $x \in X$ then $f \simeq g$. It is enough to define $\varepsilon(y) = \text{dist}(y, \mathbb{R}^n - V)$. The Corollary continues valid without assuming that either M or N is compact. We just have to approximate f by g differentiable, with $|f(x) - g(x)| < \varepsilon(f(x))$ for every x, which is possible.

Proposition 1.12. *Let $Y \subset \mathbb{R}^n$ of type ENR and $A \subset X$ be a closed subset of a normal space. Every continuous map $f\colon A \to Y$ can be extended continuously to a neighborhood of A in X.*

Proof. Let $r\colon V \to Y$ be a retraction of a neighborhood V of Y. By the Extension theorem of Tietze and Urysohn, the map f, considered as a map from A into $\mathbb{R}^n$, has a continuous extension $\varphi\colon X \to \mathbb{R}^n$. Hence, $U = \varphi^{-1}(V)$ is a neighborhood of A in X and the continuous map $\bar{f} = r \circ (\varphi|U)\colon U \to Y$ is an extension of f to the neighborhood U. □

The proposition below is the main result from this section.

Proposition 1.13. (Borsuk) *Let $Y \subset \mathbb{R}^n$ be a space of type ENR, A be a closed subset of a metric space X and $f, g\colon A \to Y$ be two homotopic continuous mappings. If f has a continuous extension $\bar{f}\colon X \to Y$ then g also admits an extension. More precisely, every homotopy between f and g extends to a homotopy between $\bar{f}$ and an extension of g.*

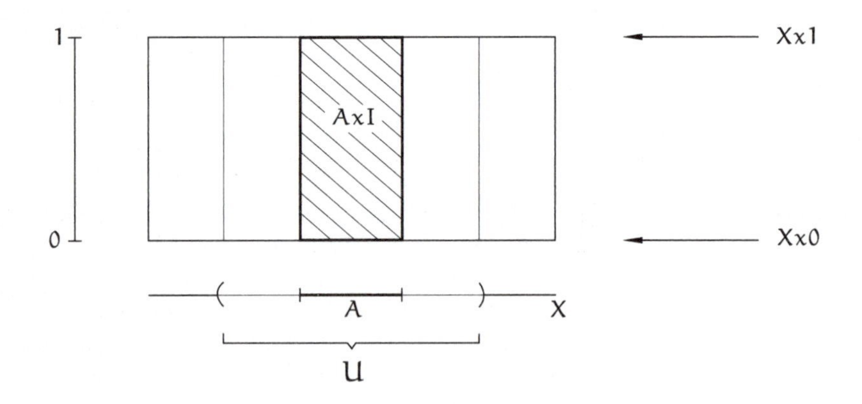

Figure 1.6.

Proof. Let $H: A \times I \to Y$ be a homotopy between f and g. Define the continuous map $f_1: (X \times 0) \cup (A \times I) \to Y$ by $f_1(x, 0) = \bar{f}(x)$ for every $x \in X$ and $f_1|(A \times I) = H$. By Proposition 1.12, f_1 extends continuously to a neighborhood W of its domain in $X \times I$. Since I is compact, there exists an open set $U \supset A$ in X such that $W \supset (X \times 0) \cup (U \times I)$. Therefore f has a continuous extension $f_2: (X \times 0) \cup (U \times I) \to Y$. Let $\lambda: X \to [0, 1]$ be a Urysohn function of the pair $(A, X - U)$; that is, f is a continuous function equal to 1 at every point of A and equal to zero in $X - U$. The extension $\bar{H}: X \times I \to Y$ of the homotopy H is then defined by $\bar{H}(x, t) = f_2(x, \lambda(x) \cdot t)$. □

Corollary 1.4. *Let $Y \subset \mathbb{R}^n$ be a contractible space of type ENR. Every continuous map $f: A \to Y$, defined on a closed subset of a metric space X, has a continuous extension $\bar{f}: X \to Y$.*

Corollary 1.5. *Let $Y \subset \mathbb{R}^n$ be a space of type ENR and $A \subset X$ be a closed contractible subset of a metric space X. Every continuous map $f: A \to Y$ admits a continuous extension $\bar{f}: X \to Y$.*

As an example, if $Y \subset \mathbb{R}^n$ is a differentiable surface or a polyhedron and $A \subset \mathbb{R}^m$ is a closed convex subset, then every continuous map $f: A \to Y$ has a continuous extension $\bar{f}: \mathbb{R}^m \to Y$.

Corollary 1.6. *Let X be a contractible metric space, $A \subset X$ be a closed subset and $Y \subset \mathbb{R}^m$ be a space of type ENR. A continuous map $f: A \to Y$ has a continuous extension $\bar{f}: X \to Y$ if, and only if, it is homotopic to a constant.*

Note that this corollary contains Proposition 1.10 in the case where the image space Y is of type ENR.

1.5 Trees

In this section, we will show that a connected graph without circuits is contractible.

An *edge* $J \subset \mathbb{R}^n$ is a subset homeomorphic to the interval $[0, 1]$. The images of the points 0 and 1 by this homeomorphism are called the *vertices* of the edge. The interior of an edge J, int. J, is called an *open edge*. Thus, an open edge is an edge without its two vertices.

A *graph* $X \subset \mathbb{R}^n$ is a union of edges that satisfies the following conditions:

1. The intersection of any two distinct edges is either empty or it consists of one or two common vertices;

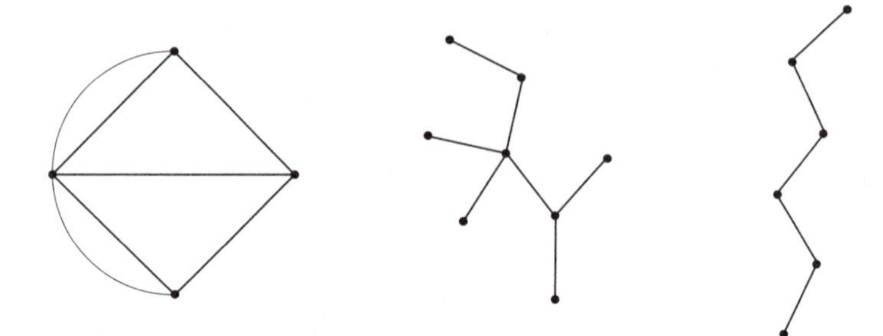

Figure 1.7. Graphs: The one on the left is the oldest and most famous of all. It was used by Leonard Euler in the problem of the Könisberg bridges. The two graphs on the right are trees.

2. A subset $S \subset X$ is closed (respectively, open) in X if, and only if, its intersection $S \cap J$ with each edge $J \subset X$ is closed (respectively, open) in the edge.

A subset $Y \subset X$ is called a *subgraph* when it is the union of edges of X. Every subgraph is a graph.

Condition 1 implies that two open edges are either disjoint or they coincide. It follows from Condition 2 that the open edges are open subsets of X and that the set of vertices is discrete and closed in X.

Let X be a graph. A map $f\colon X \to Z$, taking values on an arbitrary space Z, is continuous if, and only if, its restriction to each edge $J \subset X$ is continuous. This results immediately from Condition 2 and it is a very important remark for constructing continuous maps defined on X.

A graph is compact if, and only if, it is *finite*; that is, it has a finite number of edges (and therefore vertices). This results from the fact that the set of vertices is closed and discrete.

A graph X is connected if, and only if, any two of its vertices x, y can be joined by an *edge path*; that is, a finite sequence of edges such that x is a vertex of the first edge, y is a vertex of the last edge, and two consecutive edges of the sequence have at least a vertex in common.

A graph X is a *tree* when it is connected and $X - \text{int}.J$ is disconnected, for every edge $J \subset X$.

A *circuit* is a finite sequence $J_1, J_2, \ldots, J_n$ of distinct edges such that, indicating the vertices of J_i by u_i and v_i, we have $u_{i+1} = v_i (i = 1, 2, \ldots, n-1)$ and $v_n = u_1$. A circuit remains connected when we remove from it the interior of one of its edges. Therefore it follows that a connected graph X is a tree if, and only if, it does not contain circuits.

If X is a tree, then every connected subgraph $Y \subset X$ is also a tree.

A vertex in a graph is called *free* when it belongs to a single edge.

Lemma 1.1. *Every finite tree has a free vertex.*

Proof. Consider an edge J_1, with vertices u_1, v_1. If v_1 is not free, it will belong to another edge J_2, with vertices $u_2(= v_1)$ and v_2. In case v_2 is not free, it will belong to a new edge J_3, whose vertices we will call $u_3(= v_2)$ and v_3. And so on. Since the graph is finite, we must get to a free vertex or to an edge J_n, in which the vertex v_n coincides with one of the u_i obtained previously. This would give us a circuit, which is not possible in a tree. □

Proposition 1.14. *Every tree is contractible.*

Proof. Consider, the case of a finite tree. We use induction on the number of edges. If the tree has only one edge, it is evidently contractible. Suppose the proposition is valid for every tree with n edges and consider a tree X, with $n+1$ edges. By Lemma 1.1, there exists in X one edge J, with vertices u, v, such that v is not a vertex of any other edge of X. We can therefore deform X, contracting the edge J to the point u and leaving the other points of the tree X fixed. This gives us an homotopy equivalence between X and a subtree with n edges, which is contractible, by the induction hypothesis. Therefore X, is also contractible.

Let's consider now an infinite tree X. We construct a homotopy between the identity map and the constant map. More precisely, we take $x_0 \in X$ and define a continuous map $H\colon X \times I \to X$, such that $H(x, 0) = x$ and $H(x, 1) = x_0$ for every $x \in X$. First, we define H on the vertices and then we extend it to X.

Definition of H on the vertices: In order to define H, we take, for each vertex $v \in X$, a path $a_v\colon I \to X$ such that $a_v(0) = v$ and $a_v(1) = x_0$. (Note that the image of a_v, being compact, is contained in a finite number of edges and therefore in a finite subtree of X.) Let $V \subset X$ be the set of vertices of X. Define $H\colon V \times I \to X$ by $H(v, t) = a_v(t)$.

Extension of H to the edges: Let the vertices of J be u and v. H is already defined on the sides $\{u\} \times I, \{v\} \times I, J \times \{0\}$ and $J \times \{1\}$ of the rectangle $J \times I$, by $H(u, t) = a_u(t), H(v, t) = a_v(t), H(x, 0) = x$ and $H(x, 1) = x_0$. Let Y be a finite subtree of X containing the images by H of these four sides. Since Y is contractible, the map $H\colon \partial(J \times I) \to Y$, defined on the boundary $\partial(J \times I)$ of the rectangle $J \times I$, is homotopic to a constant and therefore, it has a continuous extension to $J \times I$, by Proposition 1.10.

Finally, taking such an extension for each edge $J \subset X$, we obtain the homotopy $H\colon X \times I \to X$ that we need. □

1.6 Homotopy of Pairs and Relative Homotopy

We say that (X, A) is a *pair* of topological spaces when A is a subspace of X.

Given the pairs (X, A) and (Y, B), a continuous map $f\colon (X, A) \to (Y, B)$ is a continuous map $f\colon X \to Y$ such that $f(A) \subset B$.

Given two continuous maps $f, g\colon (X, A) \to (Y, B)$, a *homotopy of pairs* between f and g is a continuous map

$$H\colon (X \times I, A \times I) \to (Y, B)$$

such that $H(x, 0) = f(x)$ and $H(x, 1) = g(x)$ for every $x \in X$. Therefore, we must have $H_t(A) \subset B$ for every $t \in I$.

The case where B is reduced to a point y_0 will be used frequently in the following chapters. The point y_0 is called the *base point* of the pair (Y, y_0). In this case, during a homotopy between two maps $f, g\colon (X, A) \to (Y, y_0)$, the map H_t must be constant on the subspace A.

Example 1.12. The identity map $id\colon I \to I$ is homotopic to a constant. But, considering the subspace $\partial I = \{0, 1\} \subset I$, the map of pairs $id\colon (I, \partial I) \to (I, \partial I)$ is *not* homotopic to a constant. This means that the interval I can be continuously contracted to a point but, during the deformation, at least one of its endpoints must pass through the interior of I. Indeed, any homotopy H between two maps of pairs $f, g\colon (I, \partial I) \to (I, \partial I)$ must satisfy $H_t(0) \in \partial I$ and $H_t(1) \in \partial I$ for every $t \in I$. Since $\partial I = \{0, 1\}$ is discrete, it follows that $H_t(0)$ and $H_t(1)$ do not depend of t; that is, the extreme points of I are fixed during the entire homotopy. ◁

Given two continuous maps $f, g\colon X \to Y$, we say that f is homotopic to g *relatively to a subspace* $A \subset X$, and we write $f \simeq g(\text{rel. } A)$ when there exists a homotopy $H\colon f \simeq g$ such that $H(x, t) = f(x) = g(x)$ for every $x \in A$.

Certainly, in order to have $f \simeq g$ (rel. A) it is necessary that $f(x) = g(x)$ for every $x \in A$.

Example 1.13. The identity map of $\mathbb{R}^n - \{0\}$ is homotopic to the radial projection $r\colon \mathbb{R}^n - \{0\} \to \mathbb{R}^n - \{0\}, r(x) = x/|x|$, relatively to the subspace S^{n-1}. ◁

1.7 Exercises

1. If the homeomorphisms $f,\, g\colon X \to Y$ are homotopic, their inverses f^{-1}, $g^{-1}\colon Y \to X$ are also homotopic.

2. Let $F\colon B^2 \to B^2$ be a continuous map such that $F(S^1) \subset S^1$. Define $f\colon S^1 \to S^1$ by $f(z) = F(z)$. Prove that F is surjective or f is homotopic to a constant (the two possibilities might occur).

3. An *isotopy* between the homeomorphisms $f, g\colon X \to Y$ is a homotopy $H\colon f \simeq g$ such that $H_t\colon X \to Y$ is a homeomorphism for every $t \in I$. Prove that every homeomorphism $h\colon B \to B$ of the unit ball $B = \{x \in \mathbb{R}^{n+1}; |x| \leq 1\}$, such that $h(x) = x$ for every $x \in S^n$, is isotopic to the identity map of B. Conclude that if the homeomorphisms $h, k\colon B \to B$ are defined in such a way that $h|S^n$ and $k|S^n$ are isotopic then h is isotopic to k. (Suggestion: take $H_1 =$ identity. For $0 \leq t < 1$ take $H_t =$ identity outside of the ball of center 0 and radius $1-t$ and, within this ball define H_t as a concentrated version of h.)

4. For each $\lambda \in L$, let X_λ and Y_λ be spaces with the same homotopy type. Prove that the Cartesian products

$$X = \prod_{\lambda \in L} X_\lambda \quad \text{e} \quad Y = \prod_{\lambda \in L} Y_\lambda$$

have the same homotopy type.

5. Let $E = \{(0,0,z) \in \mathbb{R}^3; z \in \mathbb{R}\}$ be the vertical axis of $\mathbb{R}^3$. Show that $\mathbb{R}^3 - E$ has the same homotopy type as the circle S^1.

6. The Möbius strip has the same homotopy type as the cylinder $S^1 \times I$ but is not homeomorphic to it.

7. Consider a compact, orientable two-dimensional surface M of genus g and a point $p \in M$. Prove that $M - \{p\}$ has the same homotopy type of the union of $2g$ circles with a point in common.

8. Let $\mathrm{GL}^+(n) \supset \mathrm{SL}(n) \supset \mathrm{SO}(n)$, respectively, the groups of $n \times n$ matrices with positive determinant, with determinant 1 and the orthogonal matrices with positive determinant. Prove that these three spaces have the same homotopy type. (Suggestion: Show that $\mathrm{GL}^+(n)$ is homeomorphic to $\mathbb{R}^+ \times \mathrm{SL}(n)$ and to $P \times \mathrm{SO}(n)$, where P is the convex set of the positive matrices.)

9. If $p < n$, $\mathbb{R}^n - \mathbb{R}^p$ has the same homotopy type of the sphere S^{n-p-1}.

10. Let X be the space obtained from the sphere S^2 by gluing the north pole to the south pole, let $Y = \mathbb{R}^3 - S^1$, where $S^1 = \{(x,y,0) \in \mathbb{R}^3; x^2+y^2 = 1\}$, and let Z be the union of a torus of revolution with a disk whose boundary is the smallest of the parallels of the torus. Prove that X, Y and Z have the same homotopy type.

11. A proper subset of S^1 is a retract of S^1 if, and only if, it is an arc of a circle.

12. A subset $Y \subset X$ is a retract of X if, and only if, every continuous map $f: Y \to Z$ has a continuous extension $\bar{f}: X \to Z$.

13. Consider an ENR space X. Prove that for every point $x_0 \in X$ and every open set $U \ni x_0$ there exists an open set V, with $x_0 \in V \subset U$ such that the mappings i, $c: V \to U$, $i(x) = x$, $c(x) = x_0$, are homotopic. (We say then that V is contractible *in* U.)

14. Consider an ENR space X and a closed subset $A \subset X$. Define an equivalence relation in X by declaring that each point $x \in X - A$ is equivalent only to itself and that the points in A are all equivalent to each other. Indicate by X/A the quotient space of X by this equivalence relation. Prove that X/A is Hausdorff. Suppose that A is contractible and prove that there exists a continuous map $f: X \to X$, homotopic to the identity, such that $f(A) =$ point. Then show that X and X/A have the same homotopy type.

15. Given a compact and convex set $C \subset \mathbb{R}^n$, we have the disjoint union $C = \mathring{C} \cup \partial C$, where the *relative interior* $\mathring{C}$ is the set of points $x \in C$ with the following property: for every $y \in C$ there exists $y' \in C$ such that x belongs to the open line segment (y, y'). The *boundary* of C is the set $\partial C = C - \mathring{C}$. Prove:

a) Every homeomorphism $h: \partial C \to \partial C$ extends to a homeomorphism $\bar{h}: C \to C$.

b) A continuous map $f: \partial C \to X$ has a continuous extension $\bar{f}: C \to X$ if, and only if, it is homotopic to a constant.

c) $\mathring{C} = \text{int}.C$ in the affine variety of $\mathbb{R}^n$ generated by C.

16. Let C be a compact and convex subset of the Euclidean space, ∂C its boundary and $\mathring{C}$ its relative interior. Take $X = C \times I$ and $Y = (C \times 0) \cup (\partial C \times I)$. Fix a point $a \in \mathring{C}$ and write $p = (a, 2) \in C \times \mathbb{R}$. Show that, for every point $(x, t) \in C \times I$, the half-line $\overrightarrow{px}$ cuts Y at a unique point $r(x)$ and that the map $r: X \to Y$, thus defined, is a retraction of $C \times I$ over $(C \times O) \cup (\partial C \times I)$. Moreover, show that r is a *deformation retraction*; that is, $i \circ r \simeq id: X \to X$, where $i: Y \to X$ is the inclusion map.

17. Let P be a polyhedron and $A \subset P$ a sub-polyhedron. Use the previous exercise to prove that $(P \times 0) \cup (A \times I)$ is a retract of $P \times I$. Conclude that, given any topological space Z and continuous mappings $\bar{f}: P \to Z$,

$g \colon A \to Z$, every homotopy between $f = \bar{f}|A$ and g extends to a homotopy between $\bar{f}$ and an extension of g. In particular, if a continuous map $A \to Z$ (with values in an arbitrary space Z) has a continuous extension to P, every map homotopic to it also admits a continuous extension.

18. Consider two metric spaces X, Y, with X compact, and the following statements:

i) The mappings $f, g \colon X \to Y$ are homotopic;

ii) For each $\varepsilon > 0$, there exist continuous mappings $f_0, f_1, \ldots, f_k \colon X \to Y$ such that $f_0 = f$, $f_k = g$, and $d(f_i(x), f_{i-1}(x)) < \varepsilon$ for every $x \in X$ and $i = 1, 2, \ldots, k$.

Prove that i) $\Rightarrow$ ii). If Y is compact of type ENR, show that the converse is also true. When X is not compact, the implication i) $\Rightarrow$ ii) may be false.

19. The Gram-Schmidt orthonormalization process provides a retraction, $r \colon \mathrm{GL}(n) \to O(n)$, from the set of $n \times n$ invertible real matrices onto the set of orthogonal matrices $n \times n$. Show that r is a deformation retraction. By restrictions of r, show that each of the spaces below is a deformation retract of the next space in the sequence:

$$\mathrm{SO}(n) \subset \mathrm{SL}(n) \subset \mathrm{GL}^+(n).$$

(We recall that a retraction $r \colon X \to Y$ is called a deformation retraction when $i \circ r \simeq id \colon X \to X$, where $i \colon Y \to X$ is the inclusion map.)

20. Prove that the following statements with respect to a set ENR, $Y \subset \mathbb{R}^n$, are equivalent:

a) Y is contractible;

b) Y is a retract of $\mathbb{R}^n$;

c) Y is a deformation retract of $\mathbb{R}^n$.

21. Prove that every finite graph G contains a maximal tree A (a subgraph that is a tree not properly contained in another tree in G). Show that a maximal tree contains every vertex of G. Use Zorn lemma, or something equivalent, to relax the hypothesis of finiteness.

22. Every finite graph has the same homotopy type of a point or of a graph without free edges.

23. Let $p = (1, 0, \ldots, 0)$ and $q = (-1, 0, \ldots, 0)$ be the north and south poles of the sphere S^n. Prove that $S^n - \{p, q\}$ has the same homotopy type as the $(n-1)$-dimensional sphere S^{n-1}.

Chapter 2

The Fundamental Group

2.1 Path Homotopy

From now on, we will consider a particular case of the general concept of homotopy, that of *path homotopy.* First, we give some preliminary definitions.

A *path* on a topological space X is a continuous map $a\colon J \to X$, defined on a compact interval $J = [s_0, s_1]$.

The points $a(s_0)$ and $a(s_1)$ are called *endpoints* of the path, $a(s_0)$ is the *initial point* (or *origin*), and $a(s_1)$ is the *final point* of the path. Geometrically, the image set $a(J)$ defines a continuous curve on the space X: as s moves from s_0 to s_1, the point $a(s)$ moves along the curve from the initial to the final point.

Let $\varphi\colon K \to J$, $K = [t_1, t_2]$ be a continuous function such that $\varphi(\partial K) \subset \partial J$. The path $b = a \circ \varphi\colon K \to X$ is called a *reparametrization* of the path a. This is illustrated by the following commutative diagram:

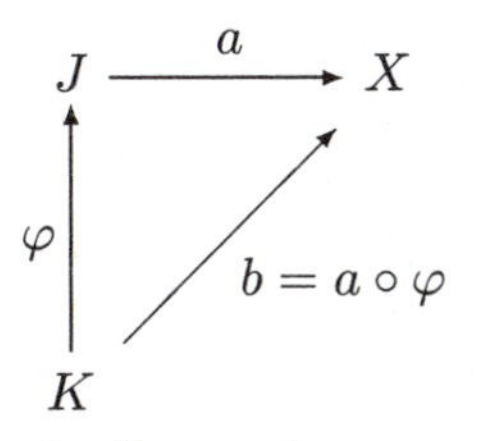

Unless we state explicitly to the contrary, we will suppose that a path is defined on the unit interval $I = [0, 1]$. This is not a major restriction because a path $a\colon [s_0, s_1] \to X$ can be reparametrized to obtain a path $a \circ \varphi\colon I \to X$, using the increasing linear homeomorphism $\varphi\colon I \to [s_0, s_1]$.

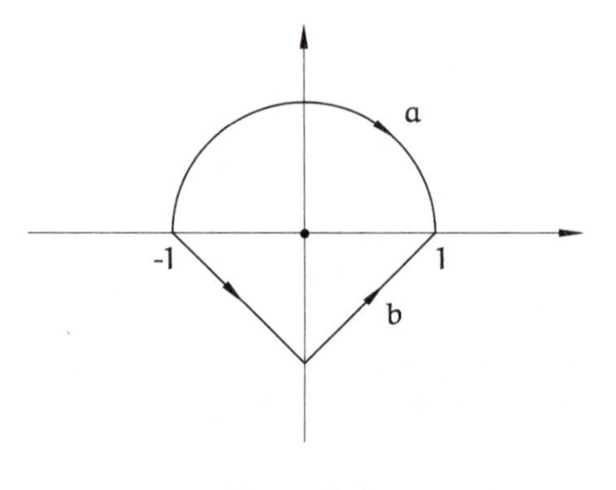

Figure 2.1.

Example 2.1. Consider the two paths $a, b\colon I \to \mathbb{R}^2 - \{0\}$, which are sketched in Figure 2.1. These paths are naturally described by the maps $a, b\colon[-1,1] \to \mathbb{R}^2 - \{0\}$, defined by $a(t) = (t, \sqrt{1-t^2})$ and $b(t) = (t, |t| - 1)$. In order to consider the paths defined on the interval $I = [0,1]$, we have to reparametrize them. A simple reparametrization is obtained using the linear homeomorphism $\varphi\colon I \to [-1,1]$, $\varphi(s) = 2s - 1$, which gives

$$(a \circ \varphi)(s) = (2s-1, \sqrt{1-(2s-1)^2}),$$

and $(b \circ \varphi)(s) = (2s-1, |2s-1| - 1)$. ◁

A path $a\colon J = [s_0, s_1] \to X$ is said to be *closed* when the two endpoints coincide; that is, $a(s_0) = a(s_1)$. We give special attention to closed paths.

Since the interval I is contractible, every path $a\colon I \to X$ is homotopic to a constant. Therefore, in order to provide substantial content to the concept of path homotopy, we must impose some restriction. We shall require that, during the homotopy, the endpoints of the path remain fixed (see Figure 2.2). This means that we will consider the boundary $\partial I = \{0,1\}$ of the interval, and the path homotopies will be relative to the subspace ∂I.

Therefore, we say that $a, b\colon I \to X$ are *homotopic paths* when $a \simeq b$ (rel. ∂I). We abbreviate this statement with the notation $a \cong b$. Thus, a homotopy $H\colon a \cong b$ between the paths a and b is a continuous map $H\colon I \times I \to X$ such that

$$\begin{aligned} &H(s,0) = a(s), H(s,1) = b(s),\\ &H(0,t) = a(0) = b(0),\\ &H(1,t) = a(1) = b(1), \end{aligned}$$

for every $s, t \in I$.

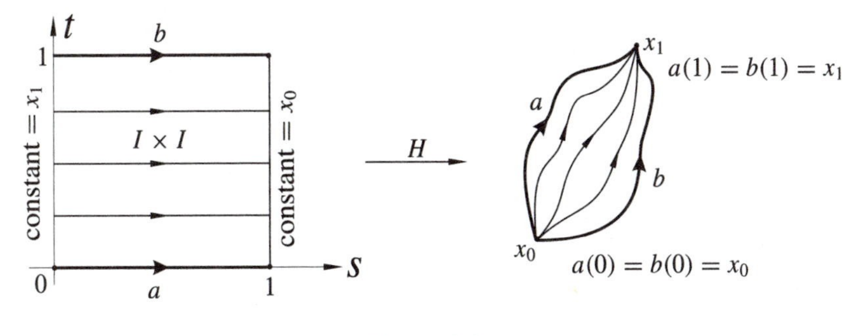

Figure 2.2.

Therefore, in order to have $a \cong b$, it is necessary that both paths a and b have the same endpoints: $a(0) = b(0) = x_0$ and $a(1) = b(1) = x_1$.

In particular, the closed paths $a, b \colon I \to X$ are homotopic (that is, $a \cong b$) when there is a continuous map $H \colon I \times I \to X$ such that, by taking $a(0) = a(1) = x_0 \in X$, we have

$$H(s,0) = a(s), \quad H(s,1) = b(s), \quad H(0,t) = H(1,t) = x_0$$

for every $s, t \in I$ (see Figure 2.3).

Example 2.2. Consider a convex subset X of a normed vector space. If $a, b \colon I \to X$ are any two paths with the same endpoints, then $a \cong b$. In fact, we just have to define $H \colon I \times I \to X$ by $H(s,t) = (1-t)a(s) + t \cdot b(s)$. It is easy to see that H is a homotopy between a and b. ◁

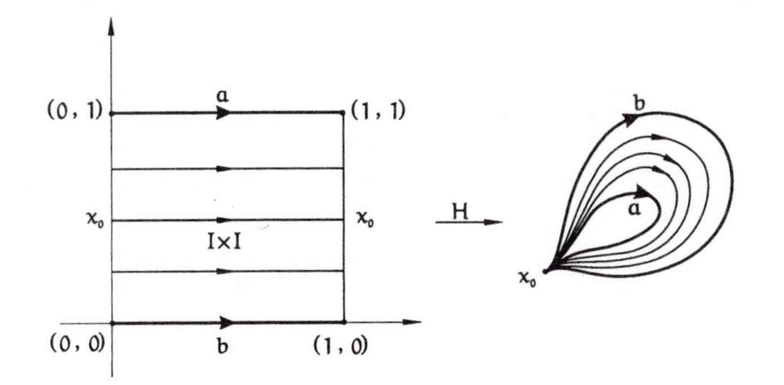

Figure 2.3.

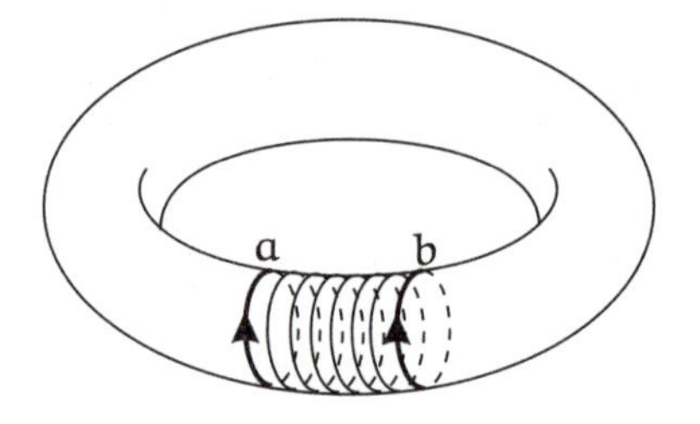

Figure 2.4. Free homotopy between two closed paths a and b on the torus.

Example 2.3. The two paths a and b in Example 2.1 are not homotopic (with fixed endpoints) in $\mathbb{R}^2 - \{0\}$. This will be proved in next chapter. Intuitively, any attempt to obtain a continuous deformation from a to b would force the deformed path to pass through the origin which does not belong to the space $\mathbb{R}^2 - \{0\}$ where the paths take their values. $\triangleleft$

Sometimes we will also consider *free homotopies* between closed paths, by allowing the endpoints to move, but requiring that the intermediate paths be closed. More precisely, two closed paths $a, b\colon I \to X$ are said to be *free homotopic* when there exists a continuous map $H\colon I \times I \to X$ such that $H(s,0) = a(s), H(s,1) = b(s)$ and $H(0,t) = H(1,t)$ for any $s, t \in I$. The last equality means that, for every $t \in I$, the path $H_t\colon I \to X, H_t(s) = H(s,t)$, is closed. Figure 2.4 shows a free homotopy between two closed paths on the two-dimensional torus.

Both path homotopy, $a \cong b$, and free homotopy are equivalence relations (that is, they are reflexive, symmetric and transitive).

We denote by $\alpha = [a]$ the homotopy class of the path $a\colon I \to X$; that is, the set of all paths in X that have the same endpoints of a and that are homotopic to a, with the endpoints fixed during the homotopy.

2.1.1 Operations with Paths

Our purpose in this chapter is to introduce a group structure associated with the space of paths. For this purpose, we introduce the product and inverse operations below.

Consider two paths $a, b\colon [s_0, s_1] \to X$ such that $a(s_1) = b(s_0)$ that is; the final point of a coincides with the origin of b (Figure 2.5). The *product* ab is the path $ab\colon [s_0, s_1] \to X$ defined by

$$ab(s) = \begin{cases} a(2s - s_0) & \text{if } s \in [s_0, \frac{s_0+s_1}{2}] \\ b(2s - s_1) & \text{if } s \in [\frac{s_0+s_1}{2}, s_1]. \end{cases}$$

Figure 2.5. The product path ab.

Geometrically, the product ab consists of first moving along the path a, and then along the path b. Since the time at our disposal to move along the path ab is equal to $s_1 - s_0$, we must double the velocity in a and in b.

Since $a(s_1) = b(s_0)$, the above definition provides a well defined map $ab\colon [s_0, s_1] \to X$, whose restrictions to the intervals $[s_0, (s_0 + s_1)/2]$ and $[(s_0 + s_1)/2, s_1]$ are continuous. It follows that ab is continuous; therefore, it is a path that starts in the initial point $a(s_0)$ of the path a, and ends in the final point $b(s_1)$ of the path b.

If $\varphi_0, \varphi_1\colon [s_0, s_1] \to [s_0, s_1]$ are the linear homeomorphisms $\varphi_0(s) = 2s - s_0$, and $\varphi_1(s) = 2s - s_1$, then the product ab coincides with the path $a \circ \varphi_0$ on the interval $[s_0, (s_0 + s_1)/2]$, and with the path $a \circ \varphi_1$ on the interval $[(s_0 + s_1)/2, s_1]$.

When the domain of both paths is I, we obtain

$$ab(s) = \begin{cases} a(2s) & \text{if } x \in [0, 1/2] \\ b(2s-1) & \text{if } x \in [1/2, 1]. \end{cases}$$

The *inverse path* of $a\colon [s_0, s_1] \to X$ is, by definition, the path $a^{-1}\colon [s_0, s_1] \to X$, given by

$$a^{-1}(s) = a(s_1 - s), \quad s \in [s_0, s_1].$$

The path a^{-1} starts at the final point and ends at the initial point of the path a. If $j\colon [s_0, s_1] \to [s_0, s_1]$ is the function $j(s) = s_1 - s$, then $a^{-1} = a \circ j$. When the path domain is I, we have $a^{-1}(s) = a(1 - s)$.

We will denote by e_x the constant path, such that $e_x(s) = x$ for every $s \in [s_0, s_1]$. Its homotopy class is denoted by $\varepsilon_x = [e_x]$.

The set of all paths in a topological space X, with the product and the inverse defined above, does not satisfy any of the group axioms. Indeed, the product ab is not defined for arbitrary pairs a, b of paths but only for the pairs that satisfy $a(1) = b(0)$.

Also, the associativity $(ab)c = a(bc)$ is not valid. In fact, consider three paths $a, b, c\colon I \to X$ such that $a(1) = b(0)$ and $b(1) = c(0)$. From the

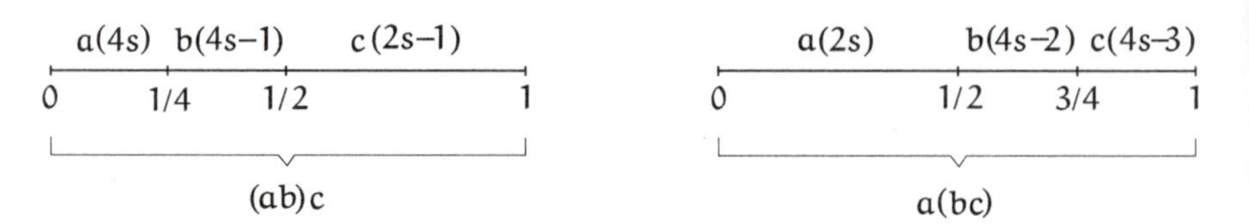

Figure 2.6.

definition of the product, we have:

$$(ab)c(s) = \begin{cases} ab(2s) & \text{if } s \in [0, 1/2] \\ c(2s-1) & \text{if } s \in [1/2, 1] \end{cases} =$$

$$= \begin{cases} a(2(2s)) & \text{if } s \in [0, 1/4] \\ b(2(2s-1/2)) & \text{if } s \in [1/4, 1/2] \\ c(2s-1) & \text{if } s \in [1/2, 1] \end{cases} = \begin{cases} a(4s) & \text{if } s \in [0, 1/4] \\ b(4s-1) & \text{if } s \in [1/4, 1/2] \\ c(2s-1) & \text{if } s \in [1/2, 1]. \end{cases}$$

A similar computation gives

$$a(bc)(s) = \begin{cases} a(2s) & \text{if } s \in [0, 1/2] \\ bc(2s-1) & \text{if } s \in [1/2, 1] \end{cases} = \begin{cases} a(2s) & \text{if } s \in [0, 1/4] \\ b(4s-2) & \text{if } s \in [1/4, 1/2] \\ c(4s-3) & \text{if } s \in [1/2, 1], \end{cases}$$

hence $(ab)c \neq a(bc)$.

The result of the above computations is summarized in the diagrams of Figure 2.6 that show how the paths $(ab)c$ and $a(bc)$ operate.

Also, there is no neutral element for path multiplication. If $a(0) = x$ and $a(1) = y$, we do not have $e_x \cdot a = a$ or $a \cdot e_y = a$. Finally, $aa^{-1} \neq e_x$ and $a^{-1} \cdot a \neq e_y$.

The desired properties for the composition law ab are found when we consider homotopy classes of paths. We will start by observing that if $a, b \colon I \to X$ are two paths satisfying $a(1) = b(0)$, then we have the following proposition

Proposition 2.1. *$a \cong a'$, $b \cong b' \Rightarrow ab \cong a'b'$, and $a^{-1} \cong (a')^{-1}$.*

Proof. If $H \colon a \cong a'$ and $K \colon b \cong b'$ are homotopies, define $L \colon I \times I \to X$ by

$$L(s,t) = \begin{cases} H(2s,t) & \text{if } 0 \le s \le 1/2, \quad t \in I \\ K(2s-1,t) & \text{if } 1/2 \le s \le 1, \quad t \in I. \end{cases}$$

Since $H(1,t) = K(0,t) = a(1) = b(0)$ for every $t \in I$, it follows that L is well defined. Since $L \mid ([0,1/2] \times I)$ and $L \mid ([1/2,1] \times I)$ are continuous, we conclude that L is continuous. Also, it is easy to prove that L is a homotopy between ab and $a'b'$. Now define $G\colon I \times I \to X$, by $G(s,t) = H(1-s,t)$. It is easy to verify that G is a homotopy between a^{-1} and $(a')^{-1}$. □

In a topological space X, consider the homotopy class α of a path that has the origin at the point $x \in X$ and the final point at $y \in X$, together with the homotopy class β of a path that starts at $y \in X$ and ends at $z \in X$. We define the product $\alpha\beta$ by taking the paths $a \in \alpha, b \in \beta$ and letting $\alpha\beta = [ab]$. Thus, by definition, $[a][b] = [ab]$. Proposition 2.1 shows that the product $\alpha\beta$ does not depend on the choices of the representative paths $a \in \alpha$ and $b \in \beta$; therefore, it is well defined.

In a similar way, we define $\alpha^{-1} = [a^{-1}]$, where $a \in \alpha$. The second part of Proposition 2.1 shows that the homotopy class $\alpha^{-1} = [a^{-1}]$ is the same for any path a that we choose in α. The class α^{-1} is called *inverse* of α.

A parametrization $\varphi\colon I \to I$ is called *positive* when $\varphi(0) = 0$ and $\varphi(1) = 1$, *negative* when $\varphi(0) = 1$ and $\varphi(1) = 0$, and *trivial* when $\varphi(0) = \varphi(1)$.

Proposition 2.2. *Let $b = a \circ \varphi$ be a reparametrization of the path $a\colon I \to X$. If the parametrization φ is positive, then $b \cong a$; if it is negative, we have $b \cong a^{-1}$; if it is trivial, then $b \cong$ constant.*

Proof. According to Example 2.2, two paths in I are homotopic (with fixed endpoints) if, and only if, they have the same initial and final points. Define $i, j\colon I \to I$ by $i(s) = s$ and $j(s) = 1 - s$. We have therefore $\varphi \cong i, \varphi \cong j$, or $\varphi \cong$ constant, provided that φ be, respectively, a positive, a negative, or a trivial reparametrization. It follows easily that $a \circ \varphi \cong a \circ i = a, a \circ \varphi \cong a \circ j = a^{-1}$, or $a \circ \varphi \cong$ constant, respectively. □

The proposition below is the result we have been looking for.

Proposition 2.3. *Consider the paths $a, b, c\colon I \to X$ such that each one of them ends where the next one starts. Let $\alpha = [a], \beta = [b]$, and $\gamma = [c]$ be their homotopy classes, $x = a(0)$ the origin of $a, y = a(1)$ its final point, e_x, e_y the constant paths on these points, and $\varepsilon_x = [e_x], \varepsilon_y = [e_y]$ the homotopy classes of these constants. We have then:*

1. $\alpha\alpha^{-1} = \varepsilon_x$;
2. $\alpha^{-1}\alpha = \varepsilon_y$;
3. $\varepsilon_x\alpha = \alpha = \alpha\varepsilon_y$;
4. $(\alpha\beta)\gamma = \alpha(\beta\gamma)$.

Proof. Consider the parametrizations $\varphi_1, \varphi_2, \varphi_3, \varphi_3', \varphi_4 \colon I \to I$, defined by:

$$\varphi_1(s) = \begin{cases} 2s & \text{if } 0 \le s \le 1/2 \\ 2 - 2s & \text{if } 1/2 \le s \le 1 \end{cases} \qquad \varphi_2(s) = \begin{cases} 1 - 2s & \text{if } 0 \le s \le 1/2 \\ 2s - 1 & \text{if } 1/2 \le s \le 1 \end{cases}$$

$$\varphi_3(s) = \begin{cases} 0 & \text{if } 0 \le s \le 1/2 \\ 2s - 1 & \text{if } 1/2 \le s \le 1 \end{cases} \qquad \varphi_3'(s) = \begin{cases} 2s & \text{if } 0 \le s \le 1/2 \\ 1 & \text{if } 1/2 \le s \le 1 \end{cases}$$

and

$$\varphi_4(s) = \begin{cases} 2s & \text{if } 0 \le s \le 1/4 \\ s + 1/4 & \text{if } 1/4 \le s \le 1/2 \\ s/2 + 1/2 & \text{if } 1/2 \le s \le 1. \end{cases}$$

Straightforward computations show that $a \circ \varphi_1 = aa^{-1}, a \circ \varphi_2 = a^{-1}a, a \circ \varphi_3 = e_x a, a \circ \varphi_3' = a e_y, (ab)c = a(bc) \circ \varphi_4$. Now, by observing that the parametrizations φ_1 and φ_2 are trivial, while φ_3, φ_3', and φ_4 are positive, the proposition follows from Proposition 2.2. □

The set of homotopy classes (with fixed endpoints) of the paths in a topological space X, with the composition law above defined, is called the *fundamental grupoid* of X and sometimes it is represented by $\Pi(X)$.

Although it is evident, it is convenient to state explicitly that, in a homotopy $a \cong bc$, the final point of b (= origin of c) is allowed to move during the process. Only the endpoints of bc (equal to those of a) must remain fixed.

2.1.2 Homotopy and Path Decomposition

We prove now that the decomposition of a path as a product of subpaths does not alter its homotopy class. This is a simple but useful result.

Consider a path $a \colon I \to X$. Take intermediate points $0 = s_0 < s_1 < \cdots < s_k = 1$, which determine a decomposition of I into k consecutive subintervals $[s_{i-1}, s_i]$. For each $i = 1, 2, \ldots, k$, we obtain a path $a_i \colon I \to X$, defined by $a_i = (a \mid [s_{i-1}, s_i]) \circ \varphi_i$, where $\varphi_i \colon I \to [s_{i-1}, s_i]$ is the unique linear function such that $\varphi_i(0) = s_{i-1}$ and $\varphi_i(1) = s_i$. The paths a_i are therefore reparametrizations of the restrictions $a \mid [s_{i-1}, s_i]$, so as to endow them with the standard domain I. Each path a_i is called a *partial path* of a.

Because $a_i(1) = a_{i+1}(0) = a(s_i)$, $i = 1, \ldots, k-1$, we may take the product $b = a_1 a_2 \ldots a_k$, which defines a path $b \colon I \to X$. Since path multiplication is not associative, it is necessary to specify the order in which the multiplication is performed. We adopt the convention $b = (((a_1 a_2) a_3) \ldots) a_k$. We remark that, by Proposition 2.3, any other distribution of parentheses in this product would give as a result a path homotopic to b (with fixed endpoints).

How does the path $b = a_1 a_2 \dots a_k$ operate?

It operates as follows: We use the points $0 = t_0 < t_1 < \dots < t_k = 1$, with $t_i = 1/2^{k-i}$, to subdivide the interval I into k consecutive subintervals $[t_{i-1}, t_i]$. For each $i = 1, 2, \dots k$, consider the increasing linear homeomorphism $\xi_i \colon [t_{i-1}, t_i] \to [s_{i-1}, s_i]$, and define a ("polygonal") increasing homeomorphism $\xi \colon I \to I$, by $\xi \mid [t_{i-1}, t_i] = \xi_i$. It follows that $b = a \circ \xi$. Therefore b is a positive reparametrization of a, hence $b \cong a$. In sum, we have the following proposition.

Proposition 2.4. *Given a path $a \colon I \to X$ and points $0 = s_0 < s_1 < \dots < s_k = 1$, consider, for each $i = 1, \dots, k, a_i \colon I \to X$ the "partial" path, defined by $a_i = (a \mid [s_{i-1}, s_i]) \circ \varphi_i$, where $\varphi_i \colon I \to [s_{i-1}, s_i]$ is the increasing linear homeomorphism. Then, by defining $b = a_1 a_2 \dots a_k$, we have $b \cong a$.*

2.2 The Fundamental Group

We will consider pairs of the type (X, x_0), where $x_0 \in X$ is called the *base point* of the topological space X. The closed path $a \colon (I, \partial I) \to (X, x_0)$ is called a closed path *based at the point* x_0. The homotopies (unless stated explicitly to the contrary) will always be relative to ∂I.

It follows from Proposition 2.3 that the subset $\pi_1(X, x_0)$ of the fundamental grupoid, which consists of the homotopy classes of closed paths based at x_0, is a group. It is called the *fundamental group* of the space X. The neutral element of this group is the homotopy class $\varepsilon = \varepsilon_{x_0}$ of the constant path at the base point x_0.

How does the choice of the base point affect the structure of the fundamental group? The answer to this question is given by the following proposition.

Proposition 2.5. *If x_0 and x_1 belong to the same pathwise connected component of X, then $\pi_1(X, x_0)$ and $\pi_1(X, x_1)$ are isomorphic. More precisely, each homotopy class γ of paths that connect x_0 to x_1 induces an isomorphism $\overline{\gamma} \colon \pi_1(X, x_1) \to \pi_1(X, x_0)$, defined by $\overline{\gamma}(\alpha) = \gamma \alpha \gamma^{-1}$.*

Proof. Let $\gamma = [c]$ be the homotopy class of a path c that connects x_0 to x_1 (see Figure 2.7). If $\alpha = [a] \in \pi_1(X, x_1)$, then $\gamma \alpha \gamma^{-1} \in \pi_1(X, x_0)$. Moreover, $\gamma(\alpha\beta)\gamma^{-1} = (\gamma\alpha\gamma^{-1})(\gamma\beta\gamma^{-1})$. Hence, $\overline{\gamma} \colon \pi_1(X, x_1) \to \pi_1(X, x_0)$, defined by $\overline{\gamma}(\alpha) = \gamma\alpha\gamma^{-1}$, is a homomorphism. Since $\alpha \mapsto \gamma^{-1}\alpha\gamma$ is a bilateral inverse to $\overline{\gamma}$, we conclude that γ is an isomorphism. □

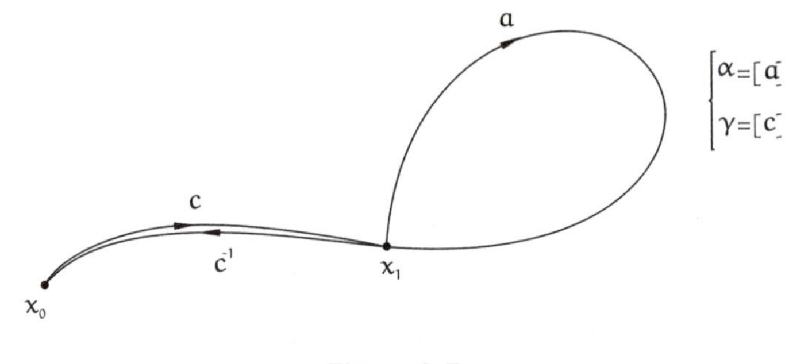

Figure 2.7.

Corollary 2.1. *If X is pathwise connected then, for any two base points $x_0, x_1 \in X$, the fundamental groups $\pi_1(X, x_0)$ and $\pi_1(X, x_1)$ are isomorphic.*

In spite of the result of Corollary 2.1, we should remark that in general, there does not exist a *natural* isomorphism (that is, one defined without arbitrary choices) between the groups $\pi_1(X, x_0)$ and $\pi_1(X, x_1)$. By changing the class γ, the isomorphism $\overline{\gamma}$ in general will also change.

The case when $\pi_1(X, x_0)$ is abelian is an exception. When this happens, any two classes γ, δ, connecting x_0 to x_1, define the same isomorphism: $\overline{\gamma} = \overline{\delta}$. In fact, in this case, for every $\alpha \in \pi_1(X, x_1)$, we have:

$$\begin{aligned}\overline{\gamma}(\alpha) = \gamma\alpha\gamma^{-1} &= (\gamma\alpha\delta^{-1})(\delta\gamma^{-1}) = \\ &= (\delta\gamma^{-1})(\gamma\alpha\delta^{-1}) = \delta\alpha\delta^{-1} = \overline{\delta}(\alpha),\end{aligned}$$

because $\gamma\alpha\delta^{-1}$ and $\delta\gamma^{-1}$ commute since they both belong to the abelian group $\pi_1(X, x_0)$.

The group $\pi_1(X, x_0)$ depends only on the pathwise connected component of the point x_0 in the space X. For this reason, it is natural to consider only pathwise connected spaces when studying the fundamental group.

2.3 The Induced Homomorphism

A continuous map $f\colon X \to Y$ induces a homomorphism

$$f_\#\colon \pi_1(X, x_0) \to \pi_1(Y, y_0), \qquad y_0 = f(x_0),$$

defined by $f_\#(\alpha) = [f \circ a]$, where $\alpha = [a]$. Since $a \cong a' \Rightarrow f \circ a \cong f \circ a'$, $f_\#$ is well defined. Moreover, it is easy to see that $f \circ (ab) = (f \circ a)(f \circ b)$, hence $f_\#(\alpha\beta) = f_\#(\alpha) f_\#(\beta)$; therefore, $f_\#$ is indeed a homomorphism.

Given two continuous maps, $f\colon X \to Y$ and $g\colon Y \to Z$, we obtain the induced homomorphisms

$$f_\#\colon \pi_1(X, x_0) \to \pi_1(Y, y_0) \quad \text{and} \quad g_\#\colon \pi_1(Y, y_0) \to \pi_1(Z, z_0),$$

with $y_0 = f(x_0)$ and $z_0 = g(y_0)$. It is easy to show that

$$(g \circ f)_\# = g_\# \circ f_\#\colon \pi_1(X, x_0) \to \pi_1(Z, z_0).$$

Finally, if $id\colon X \to X$ is the identity map, then $id_\#\colon \pi_1(X, x_0) \to \pi_1(X, x_0)$ is the identity homomorphism.

It follows from the above considerations that if $h\colon X \to Y$ is a homeomorphism, then the map $h_\#\colon \pi_1(X, x_0) \to \pi_1(Y, y_0), y_0 = h(x_0)$ is an isomorphism. Or, in a less precise form: homeomorphic spaces have isomorphic fundamental groups.

If two continuous maps $f, g\colon (X, x_0) \to (Y, y_0)$ are homotopic relative to the base point x_0 then, for every closed path $a\colon I \to X$, based at the point x_0, we have $f \circ a \cong g \circ a$; therefore, $f_\#([a]) = g_\#([a])$. Thus, f and g induce the same homomorphism between the fundamental groups.

In order to examine what happens when the image of the base point moves during the homotopy between f and g, we will study the relation between free homotopy and homotopy with a fixed base point.

Proposition 2.6. *Consider two closed paths $a, b\colon I \to X$ with bases at the points x_0, y_0 respectively. The paths a and b are free homotopic if, and only if, there exists a path $c\colon I \to X$, connecting x_0 to y_0, such that $a \cong (cb)c^{-1}$* (see Figure 2.8).

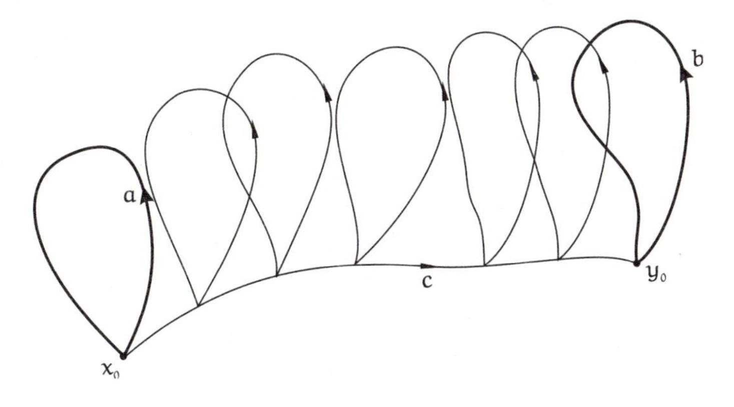

Figure 2.8.

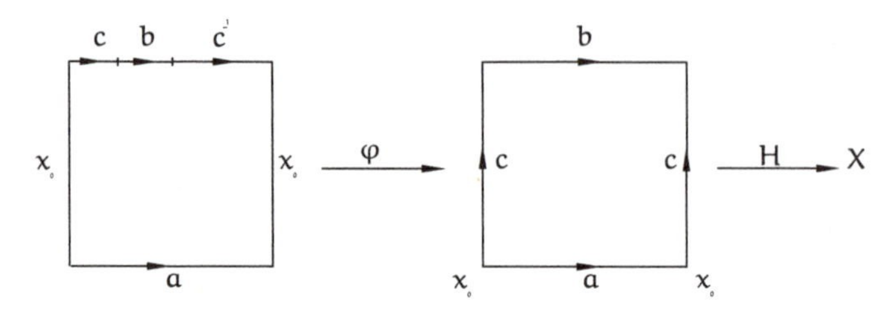

Figure 2.9.

Proof. (Necessary.) Consider a free homotopy $H\colon I\times I \to X$ between a and b. Define $c\colon I \to X$ by $c(t) = H(0,t) = H(1,t)$. Let $\varphi\colon I \times I \to I \times I$ be a continuous map that transforms the boundary of the square $I\times I$ into itself as follows (see Figure 2.9): $\varphi(0,t) = (0,0), \varphi(1,t) = (1,0), \varphi(s,0) = (s,0)$ for any $s, t \in I$; finally,

$$\varphi(s,1) = \begin{cases} (0,4s) & \text{if } 0 \le s \le 1/4 \\ (4s-1,1) & \text{if } 1/4 \le s \le 1/2 \\ (1,2-2s) & \text{if } 1/2 \le s \le 1. \end{cases}$$

The map φ does exist, because every continuous map $\dot{\varphi}\colon \partial(I\times I) \to I\times I$ from the boundary of the square to $I \times I$ extends continuously to a map φ from $I \times I$ to $I \times I$ (since $I \times I$ is contractible, $\dot{\varphi}$ is homotopic to a constant). By defining $K = H \circ \varphi$, it is easy to prove that $K\colon a \cong (cb)c^{-1}$.

(Sufficient.) Suppose $a \cong (cb)c^{-1}$. Define $H\colon I\times I \to X$ by $H = ((cb)c^{-1})\circ \varphi$, where $\varphi\colon I\times I \to I$ is a retraction from the square onto its base, as shown in Figure 2.10: φ is the identity in the base, and it transforms linearly each horizontal segment indicated in the figure into the segment $[1/4, 1/2]$ of the

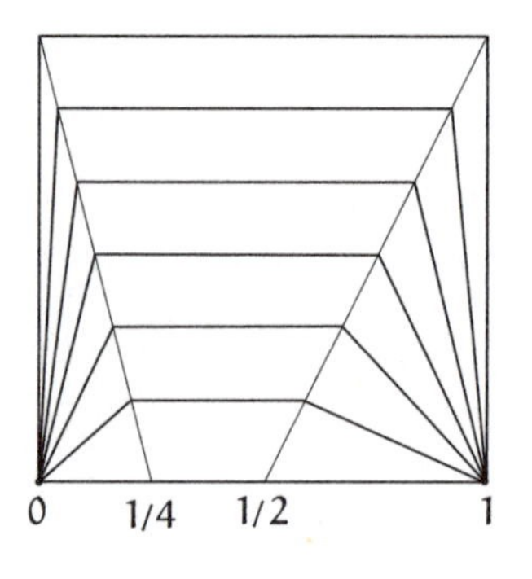

Figure 2.10.

base, each segment of origin 0 in the segment $[0, 1/4]$ and each segment with endpoint 1 in the segment $[1/2, 1]$. We have $\varphi(0,t) = t/4$, $\varphi(1,t) = 1 - t/2$ hence $H(0,t) = c(4\varphi(0,t)) = c(t)$ and $H(1,t) = c^{-1}(2\varphi(1,t) - 1) = c(t)$. Thus $H(0,t) = H(1,t)$ for every $t \in I$ and H is a free homotopy between $(cb)c^{-1}$ and b. It follows that a is free homotopic to b. □

Corollary 2.2. *If a closed path $a\colon I \to X$, based at x_0, is free homotopic to a constant then $a \cong e_{x_0}$. That is, a is homotopic to a constant without moving the point x_0 during the homotopy.*

Indeed, suppose a is free homotopic to e_{y_0}; then $a \cong (ce_{y_0})c^{-1} \cong cc^{-1} \cong e_{x_0}$.

Corollary 2.3. *Consider two closed paths $a, b\colon I \to X$ based at x_0.Let $\alpha = [a]$ and $\beta = [b]$. Then a and b are free homotopic if, and only if, α and β are conjugate elements of the group $\pi_1(X, x_0)$.*

Indeed, we just have to remark that, in this case, the path c of Proposition 2.6 is closed and based at x_0; hence, $\gamma = [c]$ belongs to $\pi_1(X, x_0)$.

From Corollary 2.3, we conclude that if a and b are closed paths with the same base point x_0, in an arbitrary topological space X, then the closed paths ab and ba are free homotopic. In fact, the homotopy classes $[ab]$ and $[ba]$ are conjugate in the group $\pi_1(X, x_0)$, since $[ab] = [b]^{-1}[ba][b]$.

We should remark, however, that not every fundamental group $\pi_1(X, x_0)$ is abelian. Examples will be given in the chapters to follow.

Corollary 2.4. *Consider two homotopic continuous maps $f, g\colon X \to Y$. The homomorphisms*

$$f_\#\colon \pi_1(X, x_0) \to \pi_1(Y, y_0) \quad \textit{and} \quad g_\#\colon \pi_1(X, x_0) \to \pi_1(Y, y_1),$$

$y_0 = f(x_0)$, $y_1 = g(x_0)$, are related by $f_\# = \overline{\gamma} \circ g_\#$, where $\overline{\gamma}\colon \pi_1(Y, y_1) \to \pi_1(Y, y_0)$ is an isomorphism defined as in Proposition 2.5. This is illustrated by the following commutative diagram:

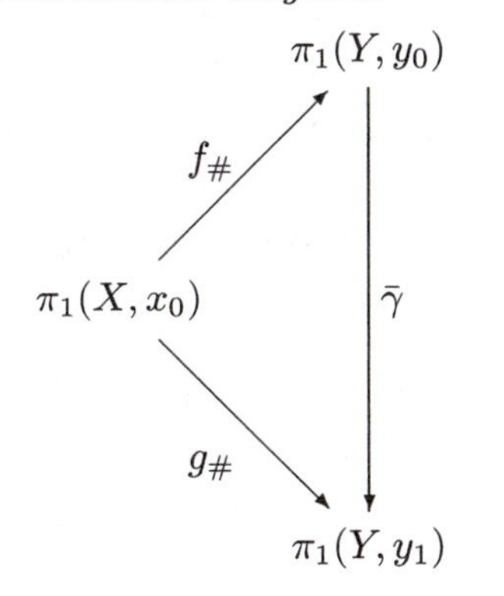

We just have to observe that if $H\colon X \times I \to Y$ is a homotopy between f and g then $c\colon I \to Y$, defined by $c(t) = H(x_0, t)$, is a path connecting $y_0 = f(x_0)$ to $y_1 = g(x_0)$. For every closed path $a\colon I \to X$, based at x_0, the map $(s,t) \mapsto H(a(s),t)$ is a free homotopy between $f \circ a$ and $g \circ a$. From Proposition 2.6, we have $f \circ a \cong c(g \circ a)c^{-1}$. By defining $\gamma = [c]$ and $\alpha = [a]$, we have: $f_\#(\alpha) = [f \circ a] = [c(g \circ a)c^{-1}] = \gamma(g_\#(\alpha))\gamma^{-1} = \overline{\gamma}(g_\#(\alpha))$.

In the following proposition, we suppose that the spaces considered are pathwise connected in order to avoid any ambiguity concerning the choice of the base points.

Proposition 2.7. *If two pathwise connected topological spaces X, Y have the same homotopy type, then their fundamental groups are isomorphic.*

Proof. Consider two continuous maps $f\colon X \to Y, g\colon Y \to X$ such that $g \circ f \simeq id_X$ and $f \circ g \simeq id_Y$. Take a base point $x_0 \in X$, and define $y_0 = f(x_0), x_1 = g(y_0), y_1 = f(x_1)$. Let $f^0_\#\colon \pi_1(X, x_0) \to \pi_1(Y, y_0)$, $f^1_\#\colon \pi_1(X, x_1) \to \pi_1(Y, y_1)$, and $g_\#\colon \pi_1(Y, y_0) \to \pi_1(X, x_1)$ be the homomorphisms induced by f and g, respectively. From Corollary 2.4, using the homotopy $g \circ f \cong id_X$, we conclude that $g_\# \circ f^0_\# = \overline{\gamma}\colon \pi_1(X, x_0) \to \pi_1(X, x_1)$, where $\overline{\gamma}$ is the isomorphism defined in Proposition 2.5 by means of the homotopy class γ of a path in X, joining x_1 to x_0. In a similar way, from $f \circ g \cong id_Y$, we conclude that $f^1_\# \circ g_\# = \overline{\delta}\colon \pi_1(Y, y_0) \to \pi_1(Y, y_1)$, where $\overline{\delta}$ is the conjugation by the homotopy class δ of a path in Y, joining y_1 to y_0. These relations express the commutativity of the diagram below.

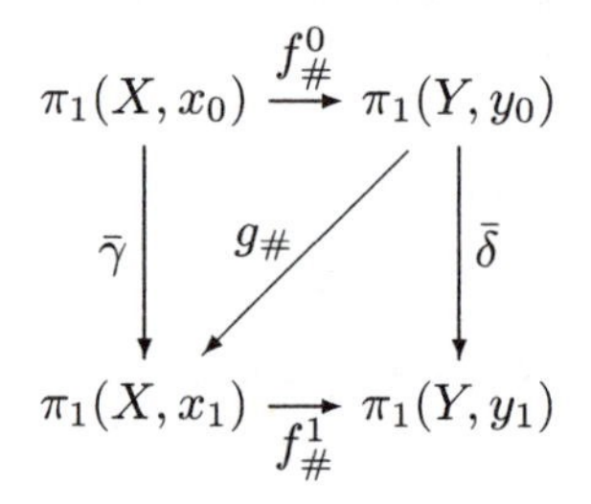

From $g_\# \circ f^0_\# = \overline{\gamma} =$ isomorphism, we conclude that $g_\#$ is surjective. From $f^1_\# \circ g_\# = \overline{\delta} =$ isomorphism, it follows that $g_\#$ is injective. Hence, $g_\#$ is an isomorphism and therefore $f^0_\#$ and $f^1_\#$ are also isomorphisms. In particular, $\pi_1(X, x_0) \approx \pi_1(Y, y_0)$. □

Corollary 2.5. *The fundamental group of a contractible space has only one element.*

We remark that this corollary also follows from Corollary 2.2.

Proposition 2.7 is very useful in practice because it allows us to perform some simplifications when computing the fundamental group. For example, the fundamental groups of $\mathbb{R}^2 - \{0\}$, $\mathbb{R}^2 - [\text{ball}]$, and S^1, are isomorphic to each other (and isomorphic to $\mathbb{Z}$, as we will show in the next chapter).

A more precise formulation of Proposition 2.7 would be: If $f\colon X \to Y$ *is a homotopy equivalence, then the induced homomorphism* $f_{\#}\colon \pi_1(X, x_0) \to \pi_1(Y, y_0), y_0 = f(x_0)$, *is an isomorphism.*

2.4 Other Descriptions of the Fundamental Group

2.4.1 Spaces with Abelian Fundamental Group

Consider a pathwise connected topological space, with $\pi_1(X, x_0)$ abelian, for some $x_0 \in X$. We know from Corollary 2.1 that, in this case, $\pi_1(X, x_1)$ is also abelian, for any choice of the base point $x_1 \in X$. Moreover, from the considerations that follow the corollary, we know that, for any two arbitrary points $x_0, x_1 \in X$, there exists a natural isomorphism $\pi_1(X, x_1) \to \pi_1(X, x_0)$; that is, to each $\alpha \in \pi_1(X, x_1)$ there corresponds a unique class $\overline{\alpha} \in \pi_1(X, x_0)$, defined without ambiguity and independent of arbitrary choices.

We will now show that, in the present case, we may consider the fundamental group of X as the set of free homotopy classes of closed paths in X and, therefore, we may represent this group by $\pi_1(X)$, without any explicit reference to the base point.

In fact, consider the set $\pi_1(X)$ of free homotopy classes of closed paths in X. Fix a base point $x_0 \in X$ and let the map $\xi\colon \pi_1(X, x_0) \to \pi_1(X)$ be such that $\xi(\alpha)$ = free homotopy class that contains α. For any pathwise connected space X, the map ξ is always surjective because every closed path a in X, based at x_1, is free homotopic to a closed path b based in x_0: we just have to take a path c connecting x_0 to x_1 and define $b = (ca)c^{-1}$. When $\pi_1(X, x_0)$ is abelian, ξ is also injective because if $\alpha = [a]$ and $\beta = [b]$ in $\pi_1(X, x_0)$ satisfy $\xi(\alpha) = \xi(\beta)$, then a and b are free homotopic. Hence α and β are conjugate elements in the commutative group $\pi_1(X, x_0)$; therefore, $\alpha = \beta$. In sum: ξ is a bijection from $\pi_1(X, x_0)$ to $\pi_1(X)$.

2.4.2 The Fundamental Group and Maps from S^1 to X

There exists a natural continuous surjection $\xi_0\colon I \to S^1$, defined by

$$\xi_0(t) = e^{2\pi i t} = (\cos 2\pi t, \sin 2\pi t).$$

Since I is compact and S^1 is Hausdorff, ξ_0 is a quotient map. This means that a map $\overline{a}\colon S^1 \to X$ is continuous if, and only if, $a = \overline{a} \circ \xi_0\colon I \to X$

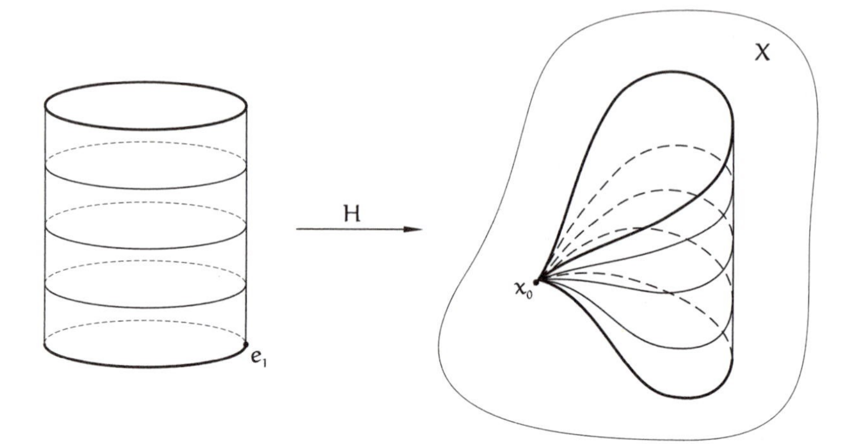

Figure 2.11.

is continuous. A path $a: I \to X$ can be written in the form $a = \overline{a} \circ \xi_0$ if, and only if, it is closed. It follows that the correspondence $\overline{a} \mapsto a = \overline{a} \circ \xi_0$ defines a bijection between the closed paths in X and the continuous maps from the circle S^1 to the space X.

Analogously, considering the continuous surjection $\zeta: I \times I \to S^1 \times I$, defined by

$$\zeta(s,t) = (\xi_0(s), t),$$

we conclude as above that there exists a bijection between the homotopies $\overline{H}: S^1 \times I \to X$ and the free homotopies of closed paths $H = \overline{H} \circ \zeta: I \times I \to X$. The closed paths $a = \overline{a} \circ \xi_0$ and $b = \overline{b} \circ \xi_0$ are free homotopic if, and only if, the corresponding maps $\overline{a}, \overline{b}: S^1 \to X$ are homotopic.

Finally, considering the point $e_1 = (1,0) \in S^1$ as a base point, with $x_0 \in X$ fixed, the closed paths $a, b: I \to X$, based at x_0, satisfy $a \cong b$ (homotopy with fixed endpoints) if, and only if, $\overline{a}, \overline{b}: (S^1, e_1) \to (X, x_0)$ are homotopic maps between pairs (or, equivalently, the image of e_1 is maintained fixed during the homotopy).

When we describe $\pi_1(X, x_0)$ as the set of homotopy classes of maps $\overline{a}: (S^1, e_1) \to (X, x_0)$, the homotopies have the cylinder $S^1 \times I$ as domain, instead of the square $I \times I$ (see Figure 2.11).

2.5 Simply Connected Spaces

A topological space X is said to be *simply connected* when one (hence all) of the following conditions hold:

1. X is pathwise connected and (for every $x_0 \in X$) we have $\pi_1(X, x_0) = \{0\}$.
2. For every closed path $a\colon I \to X$, based at the point x_0, we have $a \cong e_{x_0}$.
3. X is pathwise connected and every closed path $a\colon I \to X$ is free homotopic to a constant path (see Corollary 2.2).
4. Any two continuous maps $f, g\colon S^1 \to X$ are homotopic.
5. X is pathwise connected and every continuous map $f\colon S^1 \to X$ extends continuously to the disk $D = \{z \in \mathbb{R}^2; |z| \leq 1\}$.

Readers should convince themselves that the above statements are indeed equivalent.

Every contractible space is simply connected. In particular, the space $\mathbb{R}^n$ and balls, open and closed, are simply connected.

The necessary and sufficient condition for a pathwise connected space X to be simply connected is that any two arbitrary paths in X with the same endpoints be homotopic (with fixed endpoints). This condition is obviously sufficient because a closed path and a constant path onto its base point have the same endpoints. Conversely, if X is simply connected and $a, b\colon I \to X$ are two paths with $x_0 = a(0) = b(0)$ and $a(1) = b(1)$, then $ab^{-1} \cong e_{x_0}$, so $a \cong (ab^{-1})b \cong e_{x_0}b \cong b$.

We prove now that, when $n > 1$, the unit sphere S^n is simply connected. This follows from the lemmas below.

Remark. In the proof of the next lemma we will make use of the stereographic projection, which we describe now. Given any point $p \in S^n$, the *stereographic projection* $p\colon S^n - \{p\} \to \mathbb{R}^n$ is the homeomorphism that assigns to each $x \in S^n - \{p\}$ the point $y = \varphi(x)$, obtained as the intersection of $\mathbb{R}^n$ with the ray $\overrightarrow{px}$ that stems from p and goes through x. Here we have $S^n \subset \mathbb{R}^{n+1}$, of course, and consider $\mathbb{R}^n$ as the subset of $\mathbb{R}^{n+1}$ that consists of those points $(y_1, \ldots, y_n, 0)$ whose last coordinate is zero. If $[a, b] \subset \mathbb{R}^n$ is any line segment, then $\varphi^{-1}([a, b])$ is an arc of the circle in S^n given as the intersection $S^n \cap \Pi$, where Π is the *two*-dimensional plane in $\mathbb{R}^{n+1}$ that contains a, b, and p.

Lemma 2.1. *Let $a\colon I \to S^n$ be a path such that $a(I) \neq S^n$. Then $a \cong e_{x_0}$ if $a(0) = a(1) = x_0$ and $a \cong c$, where $c\colon I \to S^n$ is an injective path, if $a(0) \neq a(1)$.*

Proof. Take a point $p \in S^n - a(I)$. Let $\varphi\colon S^n - \{p\} \to \mathbb{R}^n$ be the stereographic projection. Since $\mathbb{R}^n$ is simply connected, $\varphi \circ a\colon I \to \mathbb{R}^n$

is homotopic (with fixed endpoints) to a constant or to a line segment (injectively parametrized), according to whether a is closed or not. The same happens with $a = \varphi^{-1} \circ (\varphi \circ a)$. □

Remarks. 1. There exist paths $a\colon I \to S^n$ (called "Peano curves" or "space filling curves") that are surjective. More precisely, a *Peano curve* in a topological space X is a path $a\colon I \to X$ such that $a(I) = X$. When $X = I \times I$ is a unit square, the first surprising example of such a path was given by the Italian mathematician Giuseppe Peano in 1890 (see Peano (1890)). A theorem of Hahn and Mazurkiewickz says that there exists a Peano curve in X if, and only if, X is a compact, connected, and locally connected Hausdorff space (see Hocking & Young (1961)).

2. The above lemma says that the only difficulty in proving that $S^n (n > 1)$ is simply connected is getting rid of the Peano curves on the sphere S^n.

3. When $a(0) \neq a(1)$, the injective path homotopic to a is indeed an arc of circle in the sphere S^n.

In the proof of Lemma 2.2 below, by an open ball B (of center x and radius r) in S^n, we mean the set $B = \{y \in S^n; |y - x| < r\}$. When $0 < r < 2$, B is homeomorphic to a ball in $\mathbb{R}^n$.

Lemma 2.2. *Let $n > 1$. If the path $a\colon I \to S^n$ is injective, its image is a closed subset with empty interior in S^n.*

Proof. Since I is compact, $a(I)$ is compact, so it is closed in S^n. Moreover, being injective, a is a homeomorphism from I onto its image $a(I)$. If $a(I)$ had a non-empty interior it would contain an open ball B, with center $x = a(s)$. We must have $B = a(J)$, where J is an open interval containing s. The path a would be a homeomorphism from J onto B, a contradiction, because $J - \{s\}$ is disconnected while $B - \{x\}$ is connected for $n > 1$. □

Lemma 2.3. *Every path $a\colon I \to S^n$ is homotopic (with fixed endpoints) to a path $b\colon I \to S^n$ such that $b(I) \neq S^n$.*

Proof. By the uniform continuity of a, we can obtain points $0 = s_0 < s_1 < \ldots < s_k = 1$ in such way that, by setting $I_i = [s_{i-1}, s_i]$, we have $a(I_i) \neq S^n$ for every $i = 1, \ldots, k$. By Proposition 2.4, we have $a \cong a_1 a_2 \ldots a_k$, where each $a_i\colon I \to S^n$ is a reparametrization of $a \mid I_i$, with $a_i(I) = a(I_i)$. From the previous lemmas, we have $a_i \cong b_i$, where the image $b_i(I)$ is a closed set with empty interior in S^n. By taking $b = b_1 b_2 \ldots b_k$, we have

$$a \cong a_1 a_2 \ldots a_k \cong b_1 b_2 \ldots b_k = b$$

and the image

$$b(I) = b_I(I) \cup \ldots \cup b_k(I)$$

is a finite union of closed sets with empty interior in S^n. It follows that $b(I)$ has empty interior. In particular, $b(I) \neq S^n$. (Note that $b(I)$ is a finite union of arcs of circles.) □

We are now able to prove the following proposition.

Proposition 2.8. *If $n > 1$, the sphere S^n is simply connected.*

Proof. By Lemma 2.3, every closed path in S^n is homotopic to a closed path whose image does not cover all of the sphere S^n. This last path, by Lemma 2.1, is homotopic to a constant. □

Example 2.4. If $n > 2$, then $\mathbb{R}^n - \{0\}$ is simply connected. In fact, $\mathbb{R}^n - \{0\}$ has the same homotopy type of the sphere S^{n-1}. ◁

Remarks. 1. The statement that S^n is simply connected when $n > 1$ is equivalent to stating that every continuous map $f\colon S^1 \to S^n$ is homotopic to a constant. More generally, it is true that for $0 < k < n$, every continuous map $f\colon S^k \to S^n$ is homotopic to a constant. Indeed, given f, we can obtain $g\colon S^k \to S^n$ of class C^1 such that $|f(x) - g(x)| < 2$ for every $x \in S^k$. Therefore $f \simeq g$ (cfr. Proposition 1.3). Now we observe that a map of class C^1 from a manifold of smaller dimension into one of greater dimension is never surjective (in fact, from Sard's theorem, its image has measure zero!). Hence, $g(S^k) \neq S^n$ and from this $g \simeq$ constant. Therefore, $f \simeq$ constant. The proof for $k = 1$, given above in Proposition 2.8, is longer but has the advantage of being elementary.

2. For $n = 1$, we have the circle S^1, which is not simply connected, as we will see in next chapter.

3. Although the sphere S^n is simply connected for $n > 1$, it is not contractible. In fact, no compact surface (without boundary) is contractible. This can be seen in many ways. If one uses homology theory, the highest dimensional homology group mod 2 of a compact surface is $\mathbb{Z}_2$, hence $\neq 0$, so it cannot be contractible. If one wishes to use analysis, one recalls that if dV is the volume element of a compact orientable surface M and $f, g\colon M \to M$ are homotopic maps, and then

$$\int_M f^* dV = \int_M g^* dV.$$

By taking $f = Id$ (identity map) and g a constant map, this would give vol. $M = 0$, a contradiction, so M is not contractible. The case of M non-orientable follows from Chapter 8 of this book, since contractible surfaces are simply connected and therefore orientable.

2.6 Some Properties of the Fundamental Group

Proposition 2.9. *The fundamental group of a Cartesian product $X \times Y$ is isomorphic to the Cartesian product of the fundamental groups of X and Y. More precisely, if $p\colon X \times Y \to X$ and $q\colon X \times Y \to Y$ are natural projections, then $\varphi\colon \pi_1(X \times Y, (x_0, y_0)) \to \pi_1(X, x_0) \times \pi_1(Y, y_0)$, defined by $\varphi(\alpha) = (p_\#(\alpha), q_\#(\alpha))$, is an isomorphism.*

Proof. A closed path $c\colon I \to X \times Y$, based at the point (x_0, y_0), has the form $c(s) = (a(s), b(s))$, where $a = p \circ c$ is a closed path in X, based in x_0, and $b = q \circ c$ is closed and based in $y_0 \in Y$. Also, given $c'\colon I \to X \times Y$, with $c'(s) = (a'(s), b'(s))$, we have $c \cong c'$ if, and only if, $a \cong a'$ and $b \cong b'$. In fact, a homotopy $K\colon c \cong c'$ can be written as $K(s,t) = (G(s,t), H(s,t))$, where $G\colon a \cong a'$ and $H\colon b \cong b'$. From this, the proposition follows. □

Corollary 2.6. *If X and Y are simply connected, then the Cartesian product $X \times Y$ is simply connected.*

If $f\colon X \to Y$ is a homeomorphism, then the induced homomorphism $f_\#$ is an isomorphism. But if $f\colon X \to Y$ is only injective, even if it is the inclusion $X \subset Y$, the homomorphism $f_\#$ may not be injective. We just have to consider, for example, the inclusion $f\colon S^1 \to \mathbb{R}^2$. Also, if $f\colon X \to Y$ is a continuous surjective map, this does not imply that the induced homomorphism $f_\#$ is surjective, as we can show by taking $f\colon \mathbb{R} \to S^1$, $f(t) = e^{it} = (\cos t, \sin t)$.

The next proposition shows that if Y is a retract of X, then the inclusion $Y \to X$ induces an injective homomorphism in the fundamental groups and if $r\colon X \to Y$ is a retraction, then the induced homomorphism $r_\#$ is surjective.

Proposition 2.10. *If $r\colon X \to Y$ is a retraction, then the induced homomorphism $r_\#\colon \pi_1(X, y_0) \to \pi_1(Y, y_0), y_0 \in Y$, is surjective and the homomorphism $i_\#\colon \pi_1(Y, y_0) \to \pi_1(X, y_0)$, induced by the inclusion map $i\colon Y \to X$, is injective. Hence, $\pi_1(Y, y_0)$ is isomorphic to a quotient group and to a subgroup $\pi_1(X, y_0)$.*

Proof. Since $r \circ i = id_Y$, we have

$$r_\# \circ i_\# = id\colon \pi_i(Y, y_0) \to \pi_1(Y, y_0).$$

It follows that $r_\#$ is surjective and that $i_\#$ is injective. □

Corollary 2.7. *If the fundamental group of X is finitely generated, the same is true for the fundamental group of any retract of X.*

Corollary 2.8. *If X is simply connected, every retract of X is also simply connected.*

Proposition 2.11. *Let $X = U \cup V$ be the union of two open pathwise connected sets, with $U \cap V$ pathwise connected. Let $i\colon U \to X, j\colon V \to X$ be the inclusion maps and $x_0 \in U \cap V$ a base point. The fundamental group $\pi_1(X, x_0)$ is generated by the images of the homomorphisms $i_\#\colon \pi_1(U, x_0) \to \pi_1(X, x_0)$ and $j_\#\colon \pi_1(V, x_0) \to \pi_1(X, x_0)$.*

Proof. Given an arbitrary closed path $a\colon I \to X$, based at x_0, we will show that $a \cong b$, where $b = b_1 b_2 \ldots b_k$ is the product of a finite number of closed paths b_i, based at x_0, such that each of them is entirely contained in U or in V. In fact, since I is compact, we obtain points $0 = s_0 < s_1 < \ldots < s_k = 1$ such that, for each $i = 1, 2, \ldots, k$, we have $a([s_{i-1}, s_i])$ contained in U or in V. If two consecutive intervals, $[s_{i-1}, s_i]$ and $[s_i, s_{i+1}]$, have both of their images contained in U or both contained in V, we eliminate the intermediate point s_i. This being done, we may suppose that $a(s_i) \in U \cap V$ for $i = 0, 1, \ldots, k$. Since $U \cap V$ is pathwise connected, we may consider paths $c_1, c_2, \ldots, c_k\colon I \to U \cap V$, such that $c_i(0) = x_0$ and $c_i(1) = a(s_i)$ (see Figure 2.12). By Proposition 2.4, we have $a \cong a_1 a_2 \ldots a_k$, where

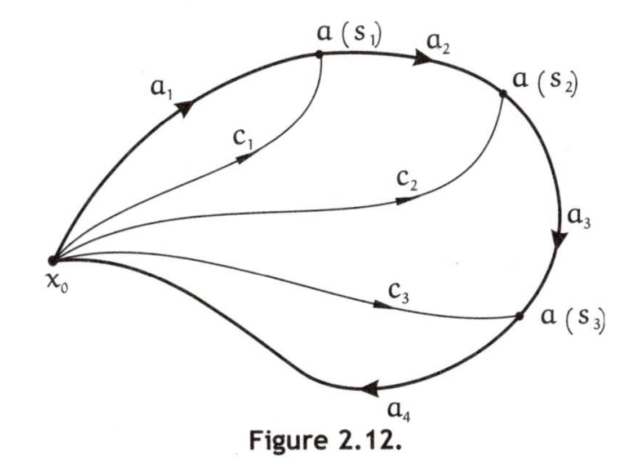

Figure 2.12.

$a_i(I) = a([s_{i-1}, s_i])$ for every i. Therefore,

$$a \cong a_1 c_1^{-1}(c_1 a_2 c_2^{-1})(c_2 a_3 c_3^{-1}) \ldots (c_k a_k) = b_1 b_2 \ldots b_k,$$

and this completes the proof. □

Corollary 2.9. *Let $X = U \cup V$, with U, V open and simply connected. If $U \cap V$ is pathwise connected, then X is simply connected.*

Remark. A more precise statement than that of Proposition 2.11 is provided by the theorem of Seifert and Van Kampen that describes $\pi_1(X)$ from the viewpoint of group theory, using the homomorphisms $i_\#$ and $j_\#$. (See references Bredon (1993), Godbillon (1971), Massey (1986) or Seifert & Threlfall (1980).)

Example 2.5. Using Corollary 2.9, we will reobtain the result that the sphere S^n of dimension $n > 1$ is simply connected. Let $p = (0, \ldots, 0, 1)$ and $q = (0, \ldots, 0, -1)$ denote the north and south poles of S^n respectively. Take $U = S^n - \{p\}$ and $V = S^n - \{q\}$. Stereographic projections show that U and V are homeomorphic to $\mathbb{R}^n$; therefore, they are simply connected. Moreover, $U \cap V = S^n - \{p, q\}$ is homeomorphic to $\mathbb{R}^n - \{0\}$ therefore it is connected for $n > 1$. Corollary 2.9 shows that $S^n = U \cup V$ is simply connected. Note that for $n = 1, U \cap V$ is disconnected, which explains why S^1 is not simply connected. ◁

Example 2.6. It is possible to apply Proposition 2.10 (and Corollary 2.9) in cases where we have $X = X_1 \cup X_2$, with $X_1 \cap X_2$ pathwise connected, without demanding that X_1 or X_2 be necessarily open in X. We just need the condition that there are open sets $U \supset X_1, V \supset X_2$ in X such that $U \cap V$ is pathwise connected and the inclusions $i \colon X_1 \to U$ and $j \colon X_2 \to V$ are homotopy equivalences. For example, let $X = S^m \cup S^n$, where S^m and S^n are spheres of dimension greater than 1, with $S^m \cap S^n = \{p\}$, a point. Take $a \in S^m$ and $b \in S^n$, both distinct from the point p, and let $U = X - \{b\}$ and $V = X - \{a\}$. The inclusions $i \colon S^m \to U$ and $j \colon S^n \to V$ are homotopy equivalences; hence, U and V are simply connected. Since $U \cap V$ is connected; Corollary 2.9 shows that $U \cup V = X$ is simply connected, that is, the union $S^m \cup S^n$ of two spheres of dimension > 1 with a point in common is simply connected. ◁

Example 2.7. It is not true, however, that the union of two simply connected spaces with a point in common is always simply connected. A counter example is given now. For each $n \in \mathbb{N}$, let Y_n and Z_n be circles of radius $1/n$ and centers at the points $(0, 0, -1/n)$ and $(0, 0, 1/n)$ respectively, in

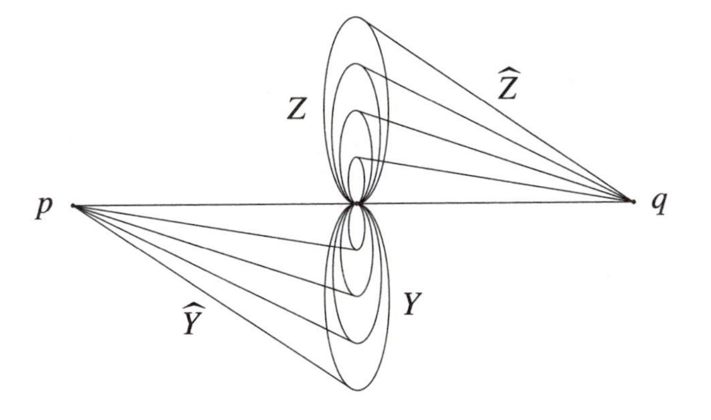

Figure 2.13.

the plane $x = 0$ in $\mathbb{R}^3$ (see Figure 2.13). Define $Y = \bigcup Y_n$, $Z = \bigcup Z_n$. We denote by $\widehat{Y}$ the cone of base Y and vertex at the point $p = (-1, 0, 0)$ in $\mathbb{R}^3$, and by $\widehat{Z}$ the cone whose vertex is the point $q = (1, 0, 0)$ and whose base is Z. Both $\widehat{Y}$ and $\widehat{Z}$ are contractible spaces (because they are stars) therefore, they are simply connected. Moreover, the intersection $\widehat{Y} \cap \widehat{Z}$ reduces to a point: the origin in $\mathbb{R}^3$. But the space $X = \widehat{Y} \cup \widehat{Z}$ is not simply connected. In fact, a closed path, based at the origin, that consists of an infinite number of circles Y_n and also an infinite number of circles Z_n is not homotopic to a constant in X. ◁

2.7 Topological Groups

A *multiplication* in a topological space X is a map $m\colon X \times X \to X$. We use the notation $m(x, y) = x \cdot y$. An element $i \in X$ is called a *neutral element* for the product when $i \cdot x = x \cdot i = x$ for every $x \in X$.

When the space X has a continuous multiplication with neutral element, it is possible to define a new composition law between paths $a, b\colon I \to X$. We define $a \cdot b\colon I \to X$ by setting

$$(a \cdot b)(s) = a(s) \cdot b(s), s \in I.$$

The path multiplication previously defined will continue to be denoted by ab.

If a and b are closed paths with bases at the neutral element $i \in X$, then the product $a \cdot b$ is a closed path based at the point i. Moreover, if the closed paths a, a', b, b' in X, based at the neutral element i, satisfy $a \cong a'$ and $b \cong b'$, then $a \cdot b \cong a' \cdot b'$.

When X is endowed with a continuous multiplication, it is possible to define a new operation between the elements of the fundamental group $\pi_1(X, i)$: Given $\alpha = [a]$ and $\beta = [b]$, define $\alpha \cdot \beta = [a \cdot b]$. From the above results, the definition of $\alpha \cdot \beta$ does not depend on the choice of the representative paths a and b in the homotopy classes of α and β.

An important particular case of a space with a continuous multiplication and neutral element is that of a *topological group* G. In this case, besides the continuous multiplication (which is associative) and the neutral element, each $x \in G$ has an inverse x^{-1}, and the map $x \mapsto x^{-1}$ is continuous.

The plane $\mathbb{R}^2$ and the unit circle $S^1 = \{(x, y) \in \mathbb{R}^2; x^2 + y^2 = 1\}$ are topological groups with respect to complex numbers multiplication. Several nontrivial examples of topological groups will be given later on, in Chapters 3 and 4.

Even when the multiplication defined on the space X is not commutative, the fundamental group $\pi_1(X, i)$, based at the neutral element i, is abelian. We will prove this now.

Lemma 2.4. *Consider a topological space X with a continuous multiplication whose neutral element is i. If $a, b\colon I \to X$ are closed paths based at i, then*

$$ae_i \cdot e_i b = ab, \quad e_i a \cdot be_i = ba,$$

where e_i is the constant path based at i.

Proof. We will use the notation s' to denote a point in the interval $[0, 1/2]$ and s'' for the interval $[1/2, 1]$. We have $ae_i(s') = a(2s')$, $ae_i(s'') = i$, while $e_i b(s') = i$, $e_i b(s'') = b(2s'' - 1)$. Therefore,

$$\begin{aligned}(ae_i \cdot e_i b)(s') &= ae_i(s') \cdot e_i b(s') = a(2s') \cdot i = a(2s') = ab(s') \\ (ae_i \cdot e_i b)(s'') &= ae_i(s'') \cdot e_i b(s'') = i \cdot b(2s'' - 1) = ab(s'').\end{aligned}$$

Hence, $ae_i \cdot e_i b = ab$. A similar argument proves the other assertion of the lemma. □

Proposition 2.12. *Let X be a topological space with continuous multiplication and neutral element i. The fundamental group $\pi_1(X, i)$ is abelian and, for arbitrary $\alpha, \beta \in \pi_1(X, i)$, the products $\alpha\beta$ and $\alpha \cdot \beta$ coincide.*

Proof. Let e_i be the constant path onto the neutral element of i. For any closed paths $a, b\colon I \to X$, based at the point i, we have

$$ab = ae_i \cdot e_i b \cong a \cdot b \cong e_i a \cdot be_i = ba.$$

This proves all of the statements of the proposition. □

Corollary 2.10. *The fundamental group of a topological group is abelian.*

2.8 Exercises

1. In a topological group G, denote temporarily by g^* the inverse of $g \in G$ and by $a^*: I \to G$ the path that we obtain from $a: I \to G$ by defining $a^*(s) = a(s)^*$. Prove that, for every closed path $a: I \to G$, based in the neutral element of G, we have $a^* \cong a^{-1}$, where $a^{-1}(s) = a(1-s)$, as usual.

2. In the metric space X, let $\mathcal{C}(X, x_0)$ be the set of paths $a, b: I \to X$ that start at the point x_0, with the metric $d(a,b) = \sup\{d(a(s), b(s)), s \in I\}$. Prove that the metric space $\mathcal{C}(X, x_0)$ is contractible.

3. Consider the subspace $\Omega(X, x_0) \subset \mathcal{C}(X, x_0)$ whose elements are the closed paths based at the point x_0. Prove that the maps $m: \Omega(X, x_0) \times \Omega(X, x_0) \to \Omega(X, x_0)$ and $i: \Omega(X, x_0) \to \Omega(X, x_0)$, defined by $m(a,b) = ab$ and $i(a) = a^{-1}$, are continuous.

4. Let C be the connected component of the point x_0 in the space X. Prove that the inclusion $i: C \to X$ induces an isomorphism $i_\#: \pi_1(C, x_0) \to \pi_1(X, x_0)$.

5. Let $r: X \to Y$ be a retraction and $i: Y \to X$ be the inclusion map. If $i_\#[\pi_1(Y, y_0)]$ is a normal subgroup of $\pi_1(X, y_0)$ (in particular, if X has abelian fundamental group), then $\pi_1(X, y_0)$ is isomorphic with the Cartesian product of the kernel of $r_\#$ by the image of $i_\#$.

6. Let $X = C_1 \cup \cdots \cup C_k$ be a finite union of convex open sets in the Euclidean space $\mathbb{R}^n$. Prove that the fundamental group of X is finitely generated. (Suggestion: In each non empty intersection $C_i \cap C_j$ choose a point $x_{ij} = x_{ji}$ and, for each pair (x_{ij}, x_{ir}) take the line segment L_{ijr} that connects them. Show that every closed path in X, based in one of the points x_{ij}, is homotopic to a polygonal path whose edges are some of the line segments L_{ijr}.)

7. The fundamental group of a polyhedron is finitely generated.

8. The fundamental group of a compact differentiable surface in $\mathbb{R}^n$ is finitely generated.

9. In Exercise 6, suppose that $C_i \cap C_j \neq \varnothing$ if, and only if, $j = i + 1$ or $\{i, j\} = \{1, k\}$. Prove that X has the homotopy type of a circle.

10. In Exercise 6, show that $\pi_1(X) = \{0\}$ if $k \leq 2$ and $\pi_1(X) = \pi_1(S^1)$ or $\pi_1(X) = \{0\}$ if $k \geq 3$.

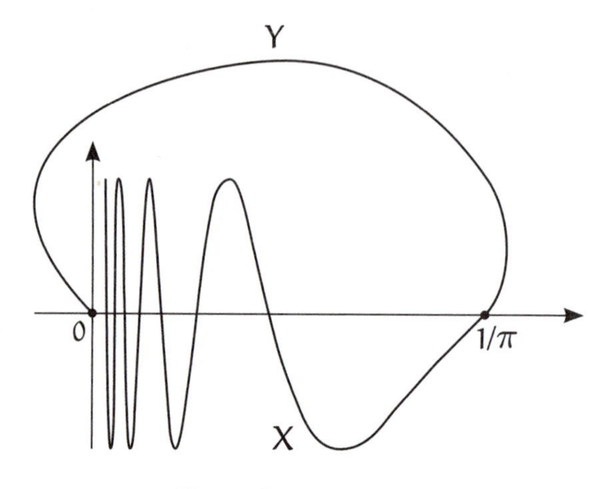

Figure 2.14.

11. Give an example of a compact topological space whose fundamental group is $\pi_1(S^1)^{\mathbb{N}}$, hence it does not have a denumerable set of generators.

12. Let $r\colon X \to Y$ be a retraction and $i\colon Y \to X$ the inclusion. Y is called a *deformation retract* of X (and r is a *retraction by deformation*) when $i \circ r \simeq id\colon X \to X$. Show that the union of a circle with one of its diameters is a deformation retract of the plane minus two points. The same happens with the union of two circles with a point in common. Conclude that these three spaces have isomorphic fundamental groups.

13. Let $f\colon [0, 1/\pi] \to \mathbb{R}$ be the function defined by $f(x) = \text{sen}(1/x)$ if $0 < x \leq 1/\pi$ and $f(0) = 0$. Represent by X the graph of f and by Y a simple arc whose endpoints are the points $(0,0)$ and $(1/\pi, 0)$, its other points have positive ordinates and do not belong to X. Define $Z = X \cup Y$ (see Figure 2.14). Prove that Z is simply connected but not locally connected.

14. Let $[X]$ be the set of homotopy classes (with fixed endpoints) of the paths in the simply connected space X. Define a bijection between $[X]$ and $X \times X$.

15. Let M be a differentiable surface of dimension n, which is a closed subset of $\mathbb{R}^k$, where $k \geq n+2$. Prove that $\mathbb{R}^k - M$ is connected. If $k \geq n+3$ then $\mathbb{R}^k - M$ is simply connected.

16. Let $X = \bigcup_{\lambda \in L} U_\lambda$ be a covering of the space X by open sets U_λ with the following properties:

a) For any arbitrary λ, $\mu \in L$, there exists $\nu \in L$ such that $U_\lambda \cup U_\mu \subset U_\nu$. (Particular case: $L = \mathbb{N}$ and $U_1 \subset \cdots \subset U_n \subset \cdots$.);

b) For each $\lambda \in L$, U_λ is pathwise connected and the homomorphism

$$i_{\lambda\#} \colon \pi_1(U_\lambda, x_\lambda) \to \pi_1(X, x_\lambda),$$

induced by the inclusion $i_\lambda \colon U_\lambda \to X$, is null. (Particular case: Every U_λ is simply connected.)

Prove that X is simply connected.

Chapter 3

Some Examples and Applications

3.1 The Fundamental Group of the Circle

We prove here that the fundamental group of the circle S^1 is infinite cyclic. This result is obtained by associating to each closed path a in the circle a number $n(a)$, called the *degree* of a, in such a way that two closed paths in the circle are homotopic if, and only if, they have the same degree. Also, every integer n is the degree of some closed path in S^1. Moreover, $n(ab) = n(a) + n(b)$, so the correspondence $a \mapsto n(a)$ induces an isomorphism between the groups $\pi_1(S^1)$ and $\mathbb{Z}$.

Initially, we observe that complex number multiplication defines a topological group structure in S^1. Hence the fundamental group of S^1 is abelian; therefore, two closed paths in S^1 with the same base point are homotopic (with the base point fixed) if, and only if, they are free homotopic.

As a starting point to compute $\pi_1(S^1)$, we look at the *exponential map*

$$\xi \colon \mathbb{R} \to S^1, \xi(t) = e^{it} = (\cos t, \sin t).$$

The equality $e^{i(s+t)} = e^{is} \cdot e^{it}$, which expresses in a simple and elegant way the classical formulas for $\cos(s+t)$ and $\sin(s+t)$, tells us that the continuous surjection ξ is a homomorphism of the additive group $\mathbb{R}$ onto the multiplicative group S^1 (complex numbers of modulus 1). The kernel of ξ is the group $2\pi\mathbb{Z} = \{2\pi n; n \in \mathbb{Z}\}$, of the integral multiples of 2π. Thus, given $u \in S^1$, we have $\xi^{-1}(u) = \{t + 2\pi n; n \in \mathbb{Z}\}$, where $t \in \mathbb{R}$ is a (any) real number such that $\xi(t) = u$. Note that $\xi(t) = u$ means that t is a determination, in radians, of the angle between u and the positive x-axis.

Remark. Sometimes we define $\xi(t) = e^{2\pi i t}$. This has the advantage of simplifying the kernel of ξ, which becomes equal to $\mathbb{Z}$. On the other hand, t does not represent any longer the measure, in radians, of the angle between $\xi(t)$ and the positive x-axis. Certainly, the two definitions are equivalent.

Lemma 3.1. *$\xi\colon \mathbb{R} \to S^1$ is an open map.*

Proof. Given an open set $U \subset \mathbb{R}$, we must prove that its image $\xi(U)$ is an open subset of S^1. It is enough to prove that the set $F = S^1 - \xi(U)$ is closed in S^1. The set $\xi^{-1}(\xi(U)) = \bigcup_{n\in\mathbb{Z}}(U + 2\pi n)$ is open in $\mathbb{R}$; therefore, its complement $\xi^{-1}(F)$ is closed in $\mathbb{R}$. Since the restriction $\xi|[0, 2\pi]$ is surjective, for each $x \in \mathbb{R}$, there exists $x' \in [0, 2\pi]$ such that $\xi(x') = \xi(x)$. Therefore $F = \xi(\xi^{-1}(F)) = \xi(\xi^{-1}(F) \cap [0, 2\pi])$. But the set $\xi^{-1}(F) \cap [0, 2\pi]$ is compact, hence its image by ξ is also compact; that is, F is compact, and therefore it is a closed subset of S^1. □

Proposition 3.1. *The restriction of ξ to any open interval $(t, t+2\pi)$ of length 2π is a homeomorphism onto $S^1 - \{\xi(t)\}$.*

Proof. The restriction $\xi|(t, t+2\pi)$ is a continuous bijection onto $S^1 - \{\xi(t)\}$. By Lemma 3.1, ξ transforms open sets in the interval $(t, t + 2\pi)$ onto open sets of S^1; therefore, the inverse of $\xi|(t, t + 2\pi)$ is also continuous. □

Corollary 3.1. *Every point $u = \xi(t) \in S^1$ has an open neighborhood $V = S^1 - \{u^*\}, u^* = -u$, whose inverse image $\xi^{-1}(V)$ is the disjoint union of open intervals $I_n = (t + \pi(2n - 1), t + \pi(2n + 1)), n \in \mathbb{Z}$, each one of which is mapped homeomorphically by ξ onto V.*

Consider a path $a\colon I \to S^1$. For each $s \in I$, since the image of ξ covers S^1, there exists some $\widetilde{s} \in \mathbb{R}$ such that $a(s) = \xi(\widetilde{s}) = e^{i\widetilde{s}}$; $\widetilde{s}$ is a determination, in radians, of the angle between $a(s)$ and the positive x-axis. The problem is that $\widetilde{s}$ is not determined in a unique way from s. We show in what follows that, for each $s \in I$, it is possible to choose $\widetilde{s}$ in such a way that $a(s) = e^{i\widetilde{s}}$ and the real map $s \mapsto \widetilde{s}$ is continuous.

Proposition 3.2. *Given an interval $J = [s_0, s_1]$, a continuous map $a\colon J \to S^1$ and a real number t_0 with $a(s_0) = e^{it_0}$, there exists a unique continuous map $\widetilde{a}\colon J \to \mathbb{R}$ such that $a(s) = e^{i\widetilde{a}(s)}$ for every $s \in J$ (that is, $a = \xi \circ \widetilde{a}$) and $\widetilde{a}(s_0) = t_0$. This is illustrated by the following commutative diagram:*

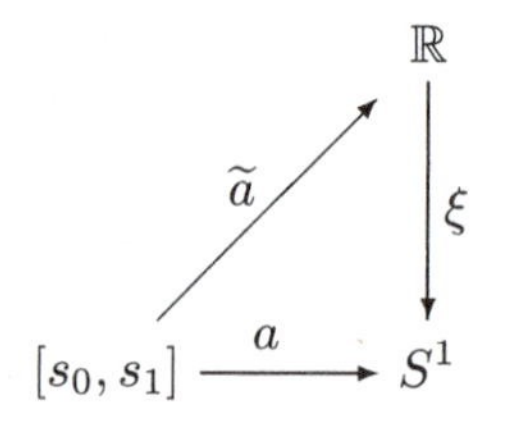

Proof. The proposition is true when $a(J) \subset S^1 - \{y\}$ for some $y \in S^1$: Since $a(s_0) \neq y$, there exists a unique $x \in \xi^{-1}(y)$ such that $t_0 \in (x, x+2\pi)$. In this case, $\xi_x = \xi|(x, x+2\pi)$ is a homeomorphism onto $S^1 - \{y\}$ and, by defining $\widetilde{a} = \xi_x^{-1} \circ a$, we obtain the desired map.

Suppose now that $J = J_1 \cup J_2$ is the union of two compact intervals with a common endpoint s_*, and that the proposition holds for the restrictions $a_1 = a|J_1$ and $a_2 = a|J_2$. We choose $\widetilde{a}_1 \colon J_1 \to \mathbb{R}$ in such a way that $\widetilde{a}_1(s_0) = t_0$ and $\xi \circ \widetilde{a}_1 = a_1$. Then we choose $\widetilde{a}_2 \colon J_2 \to \mathbb{R}$ such that $\xi \circ \widetilde{a}_2 = a_2$ and $\widetilde{a}_1(s_*) = \widetilde{a}_2(s_*)$, which is possible because $\xi(\widetilde{a}_1(s_*)) = a_1(s_*) = a_2(s_*)$. Finally, we define $\widetilde{a} \colon J \to \mathbb{R}$ by $\widetilde{a}|J_1 = \widetilde{a}_1$ and $\widetilde{a}|J_2 = \widetilde{a}_2$.

The existence of $\widetilde{a}$ in the general case follows from the two special cases above because, by the compactness of J, for every continuous map $a \colon J \to S^1$ there exists a decomposition $J = J_1 \cup \cdots \cup J_k$ as a union of consecutive intervals, in such a way that $a(J_i) \neq S^1$ for $i = 1, 2, \ldots, k$.

To prove the uniqueness, we observe that if $\widetilde{a}, \hat{a} \colon J \to \mathbb{R}$ are continuous maps such that $e^{i\widetilde{a}(s)} = e^{i\hat{a}(s)}$ for every $s \in J$, then $f(s) = [\hat{a}(s) - \widetilde{a}(s)]/2\pi$ is, for every $s \in J$, an integer that depends continuously on s. It follows that $f(s)$ is constant. In particular, if $\widetilde{a}(s_0) = \hat{a}(s_0)$, then $\widetilde{a} = \hat{a}$. □

The continuous map a in the above proposition is in fact a path on S^1 and, from the proposition, to this path is uniquely associated the path $\widetilde{a}$ in $\mathbb{R}$. We say that $\widetilde{a}$ is a *lifting* of a by the exponential map, and the exponential map has the *unique path lifting property.*

The function $\widetilde{a} \colon J \to \mathbb{R}$ is also called an *angle function* of the path a. From the above results, by fixing t_0 with $a(s_0) = \xi(t_0)$, and defining an angle function $\widetilde{a}$ with $\widetilde{a}(s_0) = t_0$, the other angle functions $\hat{a}$ for the path a, which must begin at the points $t_0 + 2k\pi, k \in \mathbb{Z}$, are related to $\widetilde{a}$ by $\hat{a}(s) = \widetilde{a}(s) + 2k\pi$.

If $a \colon I \to S^1$ is a closed path, then for every angle function $\widetilde{a} \colon I \to \mathbb{R}$ of the path a, the number

$$n(a) = \frac{\widetilde{a}(1) - \widetilde{a}(0)}{2\pi}$$

is an integer (positive, negative, or null). It does not depend on the choice of the angle function, since any two of them differ by some constant and this constant disappears when we take the difference $\widetilde{a}(1) - \widetilde{a}(0)$.

The integer $n(a)$, associated to the closed path $a \colon I \to S^1$ defined above, is called the *degree* of the path a.

The degree of a closed path $a \colon I \to S^1$ measures the "net" number of turns that the moving point $a(s)$ performs, along the road S^1, when s varies from 0 to 1. The qualification "net" means the number of positive turns (in the counterclockwise sense) minus the number of negative turns.

Proposition 3.3. *Let $a, b\colon I \to S^1$ be two closed paths. Then:*

1. *If a and b have the same base point, then $n(ab) = n(a) + n(b)$.*

2. *If a and b are free homotopic, then $n(a) = n(b)$.*

3. *If $n(a) = n(b)$, then a and b are free homotopic. Moreover, $a \cong b$ when a and b have the same base point.*

4. *For any point $p \in S^1$ and any integer $k \in \mathbb{Z}$, there exists a closed path $a\colon I \to S^1$ with a base point p, such that $n(a) = k$.*

Proof. 1. Let $\widetilde{a}, \widetilde{b}\colon I \to \mathbb{R}$ be angle functions for a and b respectively, with $\widetilde{a}(1) = \widetilde{b}(0)$. Then $\widetilde{a}\widetilde{b}\colon I \to \mathbb{R}$ makes sense and it is easy to see that $\widetilde{a}\widetilde{b}$ is an angle function for the path ab. Now, we just have to remark that $2\pi \cdot n(ab) = \widetilde{a}\widetilde{b}(1) - \widetilde{a}\widetilde{b}(0) = \widetilde{b}(1) - \widetilde{a}(0) = [\widetilde{b}(1) - \widetilde{b}(0)] + [\widetilde{a}(1) - \widetilde{a}(0)] = 2\pi[n(a) + n(b)]$.

2. Consider first the case where $|a(s) - b(s)| < 2$ for every $s \in I$; that is, the points $a(s)$ and $b(s)$ are never antipodal. In this case, we define $a(0) = e^{is_0}$ and $b(0) = e^{it_0}$. We may suppose that $|s_0 - t_0| < \pi$. Take angle functions $\widetilde{a}, \widetilde{b}$ with $\widetilde{a}(0) = s_0$ and $\widetilde{b}(0) = t_0$. Since $a(s)$ and $b(s)$ are never antipodal, we must have $\widetilde{a}(s) - \widetilde{b}(s) \neq \pi$ for every $s \in I$. This fact, along with $|\widetilde{a}(0) - \widetilde{b}(0)| < \pi$, gives us $|\widetilde{a}(s) - \widetilde{b}(s)| < \pi$ for every s. Now we have $2\pi|n(a) - n(b)| = |\widetilde{a}(1) - \widetilde{a}(0) - \widetilde{b}(1) + \widetilde{b}(0)| \leq |\widetilde{a}(1) - \widetilde{b}(1)| + |\widetilde{a}(0) - \widetilde{b}(0)| < \pi + \pi = 2\pi$. Hence, $|n(a) - n(b)| < 1$, so $n(a) = n(b)$.

The general case of any two free homotopic closed paths $a, b\colon I \to S^1$ reduces to this one. In fact, since the homotopy $H\colon I \times I \to S^1$ is uniformly continuous, there exists $\delta > 0$ such that $|t - t'| < \delta \Rightarrow |H(s,t) - H(s,t')| < 2$ for every $s \in I$. Consider $0 = t_0 < t_1 < \ldots < t_k = 1$ such that $t_{i+1} - t_i < \delta$ and define the closed paths $a_0 = a, a_1, \ldots, a_k = b$ in S^1, by $a_i(s) = H(s, t_i)$. Then $|a_i(s) - a_{i+1}(s)| < 2$ for every $s \in I$. As we have just proved, this implies that $n(a) = n(a_1) = \ldots = n(a_{k-1}) = n(b)$.

3. Consider the angle functions $\widetilde{a}, \widetilde{b}\colon I \to \mathbb{R}$ for a and b respectively. Our hypothesis $n(a) = n(b)$ assures us that $\widetilde{a}(1) - \widetilde{a}(0) = \widetilde{b}(1) - \widetilde{b}(0)$. Define a homotopy $H\colon I \times I \to \mathbb{R}$, between $\widetilde{a}$ and $\widetilde{b}$, by $H(s,t) = (1-t)\widetilde{a}(s) + t\widetilde{b}(s)$. Then, for every $t \in I$, we have $H(1,t) - H(0,t) = (1-t)[\widetilde{a}(1) - \widetilde{a}(0)] + t[\widetilde{b}(1) - \widetilde{b}(0)] = (1-t)\cdot 2\pi \cdot n + t \cdot 2\pi \cdot n = 2\pi \cdot n$, where $n = n(a) = n(b)$. From this, by taking $K = \xi \circ H$, we obtain a continuous map $K\colon I \times I \to S^1$, with $K(s,0) = a(s), K(s,1) = b(s), K(0,t) = K(1,t)$ for any $s, t \in I$. It follows that K is a free homotopy between the closed paths a and b. If a and b have the same base point, we take $\widetilde{a}(0) = \widetilde{b}(0)$, hence $\widetilde{a}(1) = \widetilde{b}(1)$, and we have $K\colon a \cong b$.

4. Let $s_0 \in \mathbb{R}$ such that $\xi(s_0) = p$. The closed path $a\colon I \to S^1$, defined by $a(s) = (\cos(s_0+2\pi ks), \sin(s_0+2\pi ks))$ is based at the point p and admits an angle function $\widetilde{a}(s) = s_0 + 2\pi k\cdot s$. Hence $n(s) = [\widetilde{a}(1) - \widetilde{a}(0)]/2\pi = k$. □

Proposition 3.3 allows us to define the degree of a homotopy class $\alpha = [a]$ of a closed path a in S^1. In fact, from item 2 of the proposition, the degree $n(a)$ depends only on the class α but not on the closed path a that we choose to represent it. Thus, we can define a map $n\colon \pi_1(S^1) \to \mathbb{Z}$ by $n(\alpha) = n(a)$.

Proposition 3.4. *The fundamental group of the circle S^1 is isomorphic to the additive group $\mathbb{Z}$ of the integers.*

Proof. Consider the map $n\colon \pi_1(S^1) \to \mathbb{Z}$ that associates to each homotopy class its degree. Item 1 of Proposition 3.3 tell us that n is a homomorphism, item 3 states that n is injective, and item 4 provides the surjectivity. Therefore, n is an isomorphism between $\pi_1(S^1)$ and $\mathbb{Z}$. □

Corollary 3.2. *The degree defines a natural bijection of the set $[S^1, S^1]$ of homotopy classes of continuous maps $f\colon S^1 \to S^1$ onto the set $\mathbb{Z}$ of integers.*

In fact, since $\pi_1(S^1)$ is abelian, there exists a natural bijection between its elements and the free homotopy classes of closed paths in S^1 which, in turn, correspond to homotopy classes of continuous maps $f\colon S^1 \to S^1$.

Corollary 3.2 may be restated by saying that each continuous map $f\colon S^1 \to S^1$ is homotopic to one, and only one, map of the type $z \mapsto z^k, k \in \mathbb{Z}$ (multiplication of complex numbers of modulus 1).

Corollary 3.3. *The fundamental group of the torus $T = S^1 \times S^1$ is free abelian with two generators.*

In fact, from Proposition 9, Chapter 2, $\pi_1(T) = \pi_1(S^1)\times\pi_1(S^1) = \mathbb{Z}\times\mathbb{Z}$.

One of the generators of $\pi_1(T)$ is the homotopy class of a parallel a and the other is the class of a meridian b. A closed path c in the torus is homotopic to $ma + nb$ if n is the net number of times that the path c cuts the parallel a and m is the net number of times that c cuts the meridian b. (Net here means that we count only when c crosses from one side to the other of a or b: it is not enough to touch. Moreover, we must count positively the crossings from one side and negatively from the other side.) For example, the path c shown in Figure 3.1 is homotopic to $a + 3b$.

The *solid torus* $S^1 \times D$ is the product of the circle S^1 by the unit closed disk $D \subset \mathbb{R}^2$. It can be represented in $\mathbb{R}^3$ by the set X, which consists

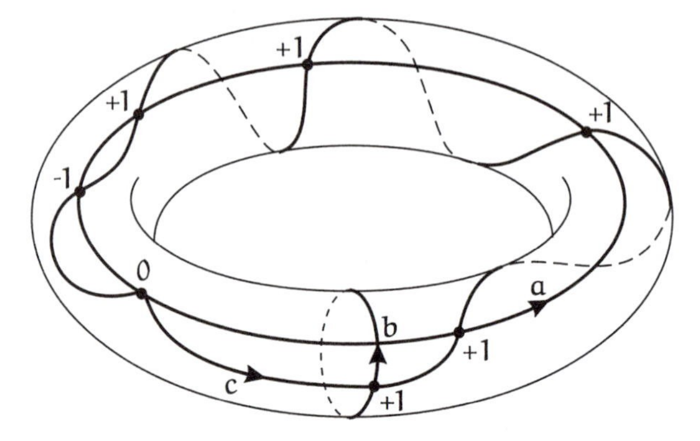

Figure 3.1. Closed path on the torus.

of the torus T with its interior. Note that the fundamental group of the solid torus is $\mathbb{Z}$, because it has the same homotopy type of the circle S^1. (See Example 1.8.) In the torus $S^1 \times S^1$, a meridian b and a parallel a are topologically indistinguishable, since the homeomorphism $(w, z) \mapsto (z, w)$ from $S^1 \times S^1$ to itself transforms the meridian $w_0 \times S^1$ on the parallel $S^1 \times w_0$. Thus, considering the subset $T \subset \mathbb{R}^3$, which gives us a geometric representation of the product $S^1 \times S^1$, there exists a homeomorphism $h\colon T \to T$ that transforms the meridian b on the parallel a. Nevertheless, we should remark that h does not extend to a homeomorphism $\overline{h}$ of the space $\mathbb{R}^3$ because $\overline{h}$ would have to take the solid torus X into itself. But, in the solid torus X, the meridian b is homotopic to a constant. Hence, it could not be transformed by $\overline{h}\colon X \to X$ in the parallel a, which defines a generator of $\pi_1(X) = \mathbb{Z}$, so it is not homotopic to a constant in X.

The nonexistence of a homeomorphism of $\mathbb{R}^3$ that transforms T onto T and takes parallels onto meridians may justify the difficulty that we have in accepting that parallels and meridians are topologically indistinguishable in the torus.

Corollary 3.4. *The fundamental group of the cylinder $C = S^1 \times \mathbb{R}$ is infinite cyclic.*

In fact, C has the same homotopy type of the circle S^1. A generator of $\pi_1(C)$ is the central circle $a(s) = (e^{is}, 0)$. A closed path c in the cylinder is homotopic to n times the generator a, where n is the net number of times that the path crosses the generating line $u \times \mathbb{R}, u = (1, 0) \in S^1$.

Next, we take a closer look at the set $[S^1, S^1]$ of homotopy classes $[f]$ of continuous maps $f\colon S^1 \to S^1$.

We recall that $[S^1, S^1]$ is in a one-to-one correspondence with $\mathbb{Z}$ as follows: To each $\varphi = [f]$, we assign the number $n(f \circ \xi_0)$, where $\xi_0 \colon I \to S^1$ is given by $\xi_0(s) = e^{2\pi i s}$. We saw that $n(f \circ \xi_0)$, the degree of the path $f \circ \xi_0 \colon I \to S^1$, depends only on φ, so we call it the *degree of* φ and write $n(f \circ \xi_0) = n(\varphi)$. The mapping $[S^1, S^1] \to \mathbb{Z}$, given by $\varphi \mapsto n(\varphi)$, is bijective.

Lemma 3.2. *Given a continuous map $f \colon S^1 \to S^1$, there exists a continuous map $\hat{f} \colon I \to \mathbb{R}$ such that the diagram below is commutative.*

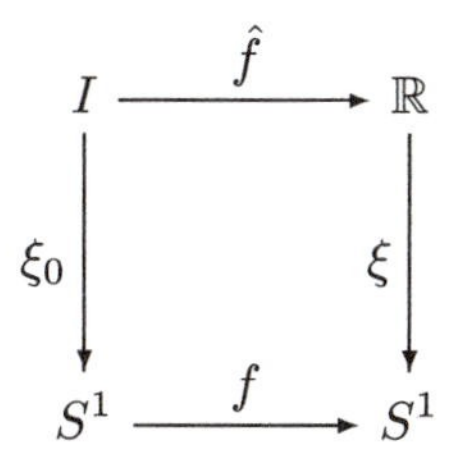

Proof. In fact, the commutativity of the diagram means that $\xi \circ \hat{f} = f \circ \xi_0$. The map $\hat{f}$ is nothing else but an angle function for the closed path $f \circ \xi_0$. $\square$

We should remark that $\hat{f}$ is determined up to an additive constant of the form $2k\pi, k \in \mathbb{Z}$ and we have $2\pi \cdot n([f]) = \hat{f}(1) - \hat{f}(0)$.

Proposition 3.5. *A continuous map $f \colon S^1 \to S^1$ is homotopic to a constant if, and only if, there exists a continuous map $\widetilde{f} \colon S^1 \to \mathbb{R}$ such that $f = \xi \circ \widetilde{f}$. This is illustrated by the commutative diagram below.*

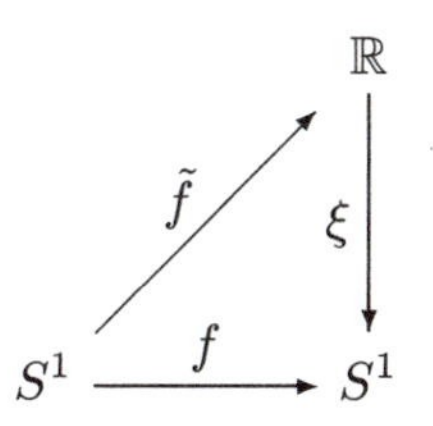

Proof. If there exists $\widetilde{f} \colon S^1 \to \mathbb{R}$ then $\widetilde{f}$ is homotopic to a constant (because $\mathbb{R}$ is contractible). It follows that $f = \xi \circ \widetilde{f}$ is also homotopic to a constant.

Conversely, suppose that f is homotopic to a constant. Then $n([f]) = 0$. Thus, the map $\hat{f}$ from Lemma 3.2 satisfies $\hat{f}(0) = \hat{f}(1)$. By passing to the quotient (see Section 2.4.2) we obtain a continuous map $\widetilde{f} \colon S^1 \to \mathbb{R}$ such that $\widetilde{f} \circ \xi_0 = \hat{f}$ therefore $\xi \circ \widetilde{f} \circ \xi_0 = \xi \circ \hat{f} = f \circ \xi_0$. Since ξ_0 is surjective, we have $\xi \circ \widetilde{f} = f$. □

3.2 The Isomorphism $\pi_1(S^1) \approx \mathbb{Z}$

In this section, we prove several statements that are consequences of the isomorphism $\pi_1(S^1) \approx \mathbb{Z}$.

C1. *S^1 is not simply connected. In particular, S^1 is not contractible.* (This is the first non trivial example that we obtain of a space with this property.) *The same happens with the torus $T = S^1 \times S^1$.*

C2. *S^1 is not a retract of the disk $D = \{(x, y) \in \mathbb{R}^2; x^2 + y^2 \leq 1\}$.*

Proof. In fact, D is simply connected; therefore, all of its retracts inherit this property. □

C3. (Brower fixed point theorem) *Every continuous map $f\colon D \to D$, from the disk into itself, has a fixed point.*

Proof. Suppose, by contradiction, that $f(z) \neq z$ for every $z \in D$. Then we define a continuous map $g\colon D \to S^1$ by

$$g(z) = \frac{f(z) - z}{|f(z) - z|}$$

for every $z \in D$. It is easy to see that g does not have fixed points. In particular, the restriction $g|S^1$ is a map from S^1 to S^1, without fixed points. Since S^1 is a sphere of odd dimension, we conclude that $g|S^1$ is homotopic to the identity $id\colon S^1 \to S^1$. (See Proposition 1.3.) Hence $n(g|S^1) = 1$. On the other hand, since $g|S^1$ extends to D, it follows that $g|S^1$ is homotopic to a constant, which gives us $n(g|S^1) = 0$, a contradiction. □

In the following two consequences (and only then), we use results from differential equations without providing additional comments.

C4. (An application of the fixed point theorem) *If a vector field v of class C^1 in the plane has a closed orbit then there exists a singular point of v in the interior of this orbit.*

Proof. The closed orbit, along with its interior region, is a set E, homeomorphic to the disk. The orbit, of each point $z \in E$ is contained in E because it cannot cross the boundary ∂E, which is another orbit.

For every $\varepsilon > 0$, let $\varphi_\varepsilon\colon E \to E$ be the continuous map that translates each point by some time ε along its orbit. If the vector field v has no singular points in E, it is possible to choose $\varepsilon > 0$ in such a way that φ_ε does not have any fixed points, but this contradicts the Brower Fixed Point theorem (Consequence C3). □

C5. (**Poincaré**) *Every continuous vector field tangent to the sphere S^2 has a singularity.*

Proof. If $v\colon S^2 \to \mathbb{R}^3$ were a continuous vector field tangent to the sphere, with $v(x) \neq 0$ for every $x \in S^2$, then there would exist $\varepsilon > 0$ such that $|v(x)| \geq 2\varepsilon$ for every $x \in S^2$. Let $u\colon S^2 \to \mathbb{R}^3$ be a vector field of class C^1 such that $|u(x) - v(x)| < \varepsilon$ for every x. Then $u(x) \neq 0$ at every point. Fix $x_0 \in S^2$. By the Poincaré-Bendixon Theorem, since there exists no singular point of u in the closure of the orbit of x_0, this orbit spirals around a closed orbit. Then, by the previous case, u has a singular point inside the closed orbit. A contradiction. □

C6. *If the continuous map $f\colon S^1 \to S^1$ is homotopic to a constant, then there exists $z \in S^1$ such that $f(-z) = f(z)$.*

Proof. By Proposition 3.5, there exists a continuous map $\widetilde{f}\colon S^1 \to \mathbb{R}$ such that $f = \xi \circ \widetilde{f}$. It is enough to prove that $\widetilde{f}$ takes the same value in some pair of antipodal points. Take $z_0 \in S^1$. If $\widetilde{f}(z_0) = \widetilde{f}(-z_0)$, we have nothing to do. Otherwise, the continuous real map $g\colon S^1 \to \mathbb{R}$, defined by $g(z) = \widetilde{f}(z) - \widetilde{f}(-z)$ takes values with opposite signs at the points z_0 and $-z_0$. Since S^1 is connected, there exists $z_1 \in S^1$ such that $g(z_1) = 0$; that is, $\widetilde{f}(z_1) = \widetilde{f}(-z_1)$. □

C7. (**Borsuk-Ulam theorem**) *For every continuous map $f\colon S^2 \to \mathbb{R}^2$, there exists $x \in S^2$ such that $f(x) = f(-x)$.*

Proof. Suppose, by contradiction, that $f(x) \neq f(-x)$ for every $x \in S^2$. The continuous map $g\colon S^2 \to S^1$, defined by the radial projection

$$g(x) = \frac{f(x) - f(-x)}{|f(x) - f(-x)|},$$

is *odd*; that is, $g(-x) = -g(x)$ for every $x \in S^2$. Restricting g to the equator

$$S^1 = \{(x_1, x_2, x_3) \in S^2; x_3 = 0\},$$

we obtain a continuous map $h = g|S^1\colon S^1 \to S^1$, homotopic to a constant (because it extends to a hemisphere containing S^1, which is a contractible set), with $h(-x) = -h(x) \neq h(x)$ for every $x \in S^1$. This contradicts Consequence C6. □

The Borsuk-Ulam theorem implies, in particular, that the sphere S^2 is not homeomorphic to a subset of the plane $\mathbb{R}^2$.

A curious interpretation of this theorem is the statement that, at each moment, there exists on the surface of the earth two antipodal points where the temperature and atmospheric pressure are equal.

Analogously, consequence C5 above says that at each moment there exists a point on the earth's surface where the wind does not blow. (Or, in other words, that we cannot comb a hairy ball.)

In Chapter 4 (see Proposition 4.4), we provide another proof of Poincaré's theorem about the tangent vector fields on S^2, this time using only results that have been proven in this book.

3.3 Real Projective Spaces

The n-dimensional *real projective space* is the quotient space $P^n = S^n/E$ of the unit sphere S^n by the equivalence relation defined as follows: Each point $x \in S^n$ is equivalent to itself or to its antipode $-x$. Each point $p \in P^n$ is therefore a non-ordered pair $p = \{x, -x\}$, $x \in S^n$.

We represent by $\pi\colon S^n \to P^n$ the natural projection, which associates to each point $x \in S^n$ its equivalence class $\pi(x) = \{x, -x\}$.

The topology of P^n is the quotient topology; that is, a set $A \subset P^n$ is (by definition) open if, and only if, its inverse image $\pi^{-1}(A)$ is an open subset of the sphere S^n.

This is the topology that we will consider in P^n. In this topology, the quotient map π is continuous and the space P^n is a Hausdorff space, which is compact, because it is the image of the compact S^n by the continuous map π.

We can express the fundamental property of the quotient topology in this case, in the following way: *If $f\colon S^n \to Y$ is a continuous map such that $f(x) = f(-x)$ for every $x \in S^n$, then there exists a unique continuous map $\overline{f}\colon P^n \to Y$ such that $\overline{f} \circ \pi = f$.*

The map $\overline{f}$ is said to be obtained from f *by passing to the quotient.* It is usual to illustrate the fundamental property with the commutative diagram below.

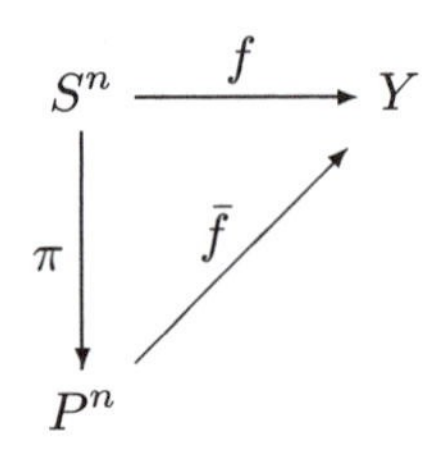

Given a subset $X \subset S^n$, we indicate by $-X$ the set of the antipodes of X. Then, for any open set $U \subset S^n$, the set $-U$ is also open, because the antipodal map is a homeomorphism.

The quotient map $\pi\colon S^n \to P^n$ is open. In fact, if $U \subset S^n$ is open, then $\pi^{-1}(\pi(U)) = U \cup (-U)$ is open in S^n; hence, $\pi(U)$ is open in P^n.

Moreover, $\pi\colon S^n \to P^n$ is locally injective. In fact, for any arbitrary point $x \in S^n$ we just have to obtain an open set $V \subset S^n$ such that $x \in V$ and $V \cap (-V) = \varnothing$. There are many choices for V; we can take, for example, $V = \{y \in S^n; \langle y, x\rangle > 0\}$. Then $\pi|V$ is injective.

Using the last two remarks, we conclude that every point of S^n has an open neighborhood which is mapped homeomorphically by $\pi\colon S^n \to P^n$ onto an open set of P^n. A more precise statement is provided by the following proposition.

Proposition 3.6. *Every point $p \in P^n$ has an open neighborhood V whose inverse image $\pi^{-1}(V) = \widetilde{V} \cup (-\widetilde{V})$ is the union of two open sets, each one of them is mapped by π homeomorphically onto V.*

Proof. Let $p = \{x, -x\} \in P^n$. Take an open neighborhood $\widetilde{V} \ni x$ in S^n such that $\widetilde{V} \cap (-\widetilde{V}) = \varnothing$; that is, $\widetilde{V}$ does not contain the antipode of any of its points. Since π is open, the set $V = \pi(\widetilde{V})$ is an open neighborhood of p, with $\pi^{-1}(V) = \widetilde{V} \cup (-\widetilde{V})$ and the restrictions $\pi|\widetilde{V}$, $\pi|(-\widetilde{V})$ are homeomorphisms onto V. □

The neighborhood V with the above property is called a *distinguished neighborhood* of the point $p \in P^n$.

As in Proposition 3.2, we now show that the quotient map $\pi\colon S^n \to P^n$ has the unique path lifting property.

Proposition 3.7. *Consider a path $a\colon [s_0, s_1] \to P^n$ and a point $x_0 \in S^n$ such that $\pi(x_0) = a(s_0)$. There exists a unique path $\widetilde{a}\colon [s_0, s_1] \to S^n$ such that $\widetilde{a}(s_0) = x_0$ and $a = \pi \circ \widetilde{a}$ (*see Figure 3.2*).*

Proof. The result is valid in the particular case when the image of the path a is contained in a distinguished neighborhood $V \subset P^n$, with $\pi^{-1}(V) = \widetilde{V} \cup (-\widetilde{V})$. In fact, we choose the notation in such a way that $x_0 \in \widetilde{V}$. Let $h = \pi|\widetilde{V}$. Then h is a homeomorphism from $\widetilde{V}$ to V. Now take $\widetilde{a} = h^{-1} \circ a\colon [s_0, s_1] \to \widetilde{V}$.

We now consider another particular case, in which the interval $J = [s_0, s_1]$ can be subdivided as the union $J = J_1 \cup J_2$ of two compact intervals with a common endpoint s_* and that the proposition is valid for the

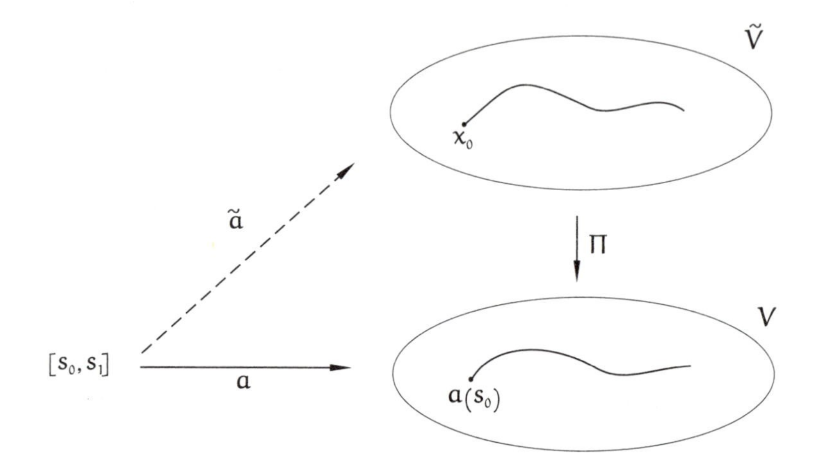

Figure 3.2. Path lifting property of the quotient map.

restrictions $a|J_1 = a_1$ and $a|J_2 = a_2$. We obtain $\widetilde{a}_1 \colon J_1 \to S^n$ such that $\widetilde{a}(s_0) = x_0$ and $\pi \circ \widetilde{a}_1 = a_1$, and $\widetilde{a}_2 \colon J_2 \to S^n$ such that $\pi \circ \widetilde{a}_2 = a_2$ and $\widetilde{a}_2(s_*) = \widetilde{a}_1(s_*)$. Now we define $\widetilde{a} \colon J \to S^n$ by $\widetilde{a}|J_1 = \widetilde{a}_1$ and $\widetilde{a}|J_2 = \widetilde{a}_2$.

The existence of $\widetilde{a}$ in the general case reduces to these two particular cases. In fact, given $a \colon J \to P^n$, the compactness of J provides a decomposition $J = J_1 \cup \cdots \cup J_k$ of J into compact consecutive intervals such that, for every $i = 1, 2, \ldots, k$, $a(J_i) \subset V_i$, where V_i is a distinguished neighborhood.

The uniqueness of $\widetilde{a}$ can be proved as follows: Suppose that $\widetilde{a}, \hat{a} \colon J \to S^n$ are two paths such that $\pi \circ \widetilde{a} = \pi \circ \hat{a}$. Then, for every $s \in J$, we must have $\widetilde{a}(s) = \hat{a}(s)$ or $\widetilde{a}(s) = -\hat{a}(s)$. Taking the inner product, we obtain $\langle \widetilde{a}(s), \hat{a}(s) \rangle = \pm 1$ for every $s \in J$. Since J is connected, this inner product must be constant. If $\widetilde{a}(s_0) = \hat{a}(s_0)$ then we must have $\widetilde{a}(s) = \hat{a}(s)$ for all $s \in J$. □

The path $\widetilde{a}$ is called a *lifting* of the path a. Note that, given $a \colon J \to P^n$, there exist precisely two liftings $\widetilde{a}, \hat{a} \colon J \to S^n$. We have $\widetilde{a}(s) = -\hat{a}(s)$ for every $s \in J$.

We now show that the topology of P^n can be defined using a very simple and natural metric.

Define the distance between the points $p = \{x, -x\}$ and $q = \{y, -y\}$ in P^n, by:

$$d(p, q) = \min\{|x - y|, |x + y|\}.$$

Geometrically, $d(p, q)$ is the smaller side of the rectangle whose vertices are $x, -x, y, -y$ (see Figure 3.3). The reader may verify that, thus defined, the

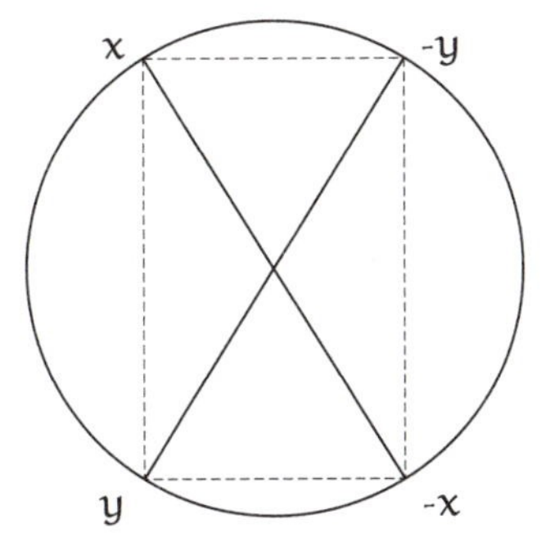

Figure 3.3.

distance $d(p,q)$ satisfies the axioms of a metric space. We use the notation (P^n, d) to represent the real projective space with the metric d. We still denote the quotient topological space $P^n = S^n/E$ by P^n.

Proposition 3.8. *The metric d, introduced above, defines in P^n its usual topology.*

Proof. The map $\pi \colon S^n \to (P^n, d)$ satisfies the condition $d(\pi(x), \pi(y)) \leq |x - y|$; hence, it is continuous. Passing to the quotient, we obtain the continuous map $\overline{\pi} \colon P^n \to (P^n, d)$, which is nothing more than the identity map. Since P^n is compact and (P^n, d) is Hausdorff, we conclude that $\overline{\pi}$ is a homeomorphism; that is, d defines in P^n its own topology. □

The metric space (P^n, d) has diameter equal to $\sqrt{2}$. In fact, the greatest value of the distance $d(p,q)$ between the points $p = \{x, -x\}$ and $q = \{y, -y\}$ is attained when the rectangle of vertices $\{x, -x, y, -y\}$ is a square. In this case, we have $|x - y| = |x + y| = \sqrt{2}$; therefore, $d(p,q) = \sqrt{2}$. When this happens, we say that p and q are *opposite points* in P^n. We remark that if $d(p,q) = \sqrt{2}$ then $x \in \pi^{-1}(p)$, $y \in \pi^{-1}(q) \Rightarrow |x - y| = \sqrt{2}$.

Now we compute the fundamental group of the projective space P^n. We just have to consider the cases where $n \geq 2$ because the case $n = 1$ is singular, as we show next.

Proposition 3.9. *The projective space P^1 is homeomorphic to the circle S^1.*

Proof. The continuous map $f \colon S^1 \to S^1$, given by $f(z) = z^2$, (multiplication of complex numbers) is such that $f(z) = f(w) \Leftrightarrow w = \pm z$. Hence, by passing to the quotient, f induces a continuous bijection $\overline{f} \colon P^1 \to S^1$ such that $\overline{f} \circ \pi = f$. Since P^1 is compact and S^1 is Hausdorff, it follows that the continuous bijection $\overline{f}$ is a homeomorphism of P^1 onto S^1. □

From the above result, we have $\pi_1(P^1) \approx \mathbb{Z}$. Now we examine $\pi_1(P^n)$ for $n \geq 2$.

Our analysis is based in the following remark: The lifting $\widetilde{a}: I \to S^n$ of a closed path $a: I \to P^n$ is not always a closed path. (We should recall that a has two liftings, $\widetilde{a}$ and $-\widetilde{a}$. One of them is closed if, and only if, both are closed.)

In order to clarify this point, consider a path $\widetilde{a}: I \to S^n$ on the sphere. Let $a = \pi \circ \widetilde{a}: I \to P^n$ be its projection in the projective space P^n. Then $\widetilde{a}$ is a lifting of a. If $\widetilde{a}$ is closed, the closed path a has a closed lifting. However, if the endpoints of $\widetilde{a}$ are antipodal points, a is a closed path in P^n whose lifting is not closed.

Proposition 3.10. *For $n \geq 2$, let $x_0 \in S^n$ and $p_0 = \pi(x_0) \in P^n$. Consider two closed paths $a, b: I \to P^n$, with base p_0, and let $\widetilde{a}, \widetilde{b}: I \to S^n$ be their liftings with origin x_0. Then $\widetilde{a}(1) = \widetilde{b}(1)$ if, and only if, $a \cong b$.*

Proof. If $\widetilde{a}(1) = \widetilde{b}(1)$, then $\widetilde{a}$ and $\widetilde{b}$ are paths in S^n with the same origin and the same endpoints. Since S^n is simply connected, we have $\widetilde{a} \cong \widetilde{b}$. It follows that $a = \pi \circ \widetilde{a} \cong \pi \circ \widetilde{b} = b$.

To prove the converse, suppose that $a \cong b$. Note that $\widetilde{a}(0) = \widetilde{b}(0) = x_0$ and $\widetilde{a}(1) = \pm x_0$, $\widetilde{b}(1) = \pm x_0$. Thus, $\widetilde{a}(1) = \widetilde{b}(1) \Leftrightarrow |\widetilde{a}(1) - \widetilde{b}(1)| \neq 2$.

First, we consider the particular case in which $d(a(s), b(s)) \neq \sqrt{2}$ for every $s \in I$, that is, the points $a(s)$ and $b(s)$ are never opposite. Then $|\widetilde{a}(s) - \widetilde{b}(s)| \neq \sqrt{2}$ for every s, which prevents that we have $|\widetilde{a}(1) - \widetilde{b}(1)| = 2$, because $\widetilde{a}(0) = \widetilde{b}(0)$.

Now we prove the general case. Given a homotopy $H: I \times I \to P^n$ between a and b, its uniform continuity yields the existence of points $0 = t_0 < t_1 < \cdots < t_k = 1$ such that $d(H(s, t_{i-1}), H(s, t_1)) < \sqrt{2}$ for every $s \in I$, $i = 1, 2 \ldots, k$. Define the closed paths $a_0, a_1, \ldots, a_k$ with base p_0, by $a_i(s) = H(s, t_i)$. From the particular case that we have proved above, we have $\widetilde{a}_i(1) = \widetilde{a}_{i+1}(1)$, because the points $a_i(s)$ and $a_{i+1}(s)$ are never opposite. Hence $\widetilde{a}(1) = \widetilde{a}_0(1) = \widetilde{a}_1(1) = \cdots = \widetilde{a}_k(1) = \widetilde{b}(1)$. □

Corollary 3.5. *For $n \geq 2$, the fundamental group of the real projective space P^n has two elements; therefore, it is isomorphic to $\mathbb{Z}_2$.*

In fact, there are only two classes of homotopy of closed paths in P^n with a base at the point p_0: the class of the paths whose lifting is closed and those which have an open lifting.

In order to exhibit explicitly the generator of the group $\pi_1(P^n)$, let $e_1 = (1, 0, \ldots, 0)$, $e_2 = (0, 1, 0, \ldots, 0)$ and consider the path $\widetilde{a}: I \to S^n$, $\widetilde{a}(s) = \cos \pi s \cdot e_1 + \sin \pi s \cdot e_2$. The image of this path is an arc of great circle connecting the antipodal points e_1, and $-e_1$ of S^n. The closed path $a = \pi \circ \widetilde{a}$ in P^n is not homotopic to a constant; hence, its homotopy class $\alpha = [a]$ generates $\pi_1(P^n)$. We have $2\alpha = 0$ in $\pi_1(P^n)$ and this reflects the fact that the path aa has a closed lifting in S^n.

The reader may have noticed the analogy between the methods used in this chapter to determine the fundamental groups of the circle S^1 and of the projective space $P^m (m \geq 2)$. The essential instruments were the maps $\xi\colon \mathbb{R} \to S^1$ (exponential map) and $\pi\colon S^m \to P^m$ (quotient map). Both are local homeomorphisms with simply connected domains and with the property of unique path lifting. ξ and π are examples of universal coverings.

The general theory of covering spaces will be studied in the second part this book. The particular cases that we have just considered with ad hoc arguments are useful as motivation for this future study.

In the remaining examples of this chapter we use rather different techniques. Instead of covering spaces, we use the more general notion of a locally trivial fibration, which constitutes one of the basilar concepts of topology and its different applications.

3.4 Fibrations and Complex Projective Spaces

The (real) unit sphere S^{2n+1} may be considered as a set of $(n+1)$-lists of complex numbers $z = (z_1, z_2, \ldots, z_{n+1})$ such that

$$|z_1|^2 + \cdots + |z_{n+1}|^2 = 1.$$

The multiplicative group S^1 of the complex numbers of modulus 1 acts on S^{2n+1} in a natural way: For each $u \in S^1$ and each $z \in S^{2n+1}$, we define

$$u \cdot z = (u \cdot z_1, \ldots, u \cdot z_{n+1}) \in S^{2n+1}.$$

The *orbit* of a point $z \in S^{2n+1}$ with respect to this action of S^1 is the set $\{u \cdot z; u \in S^1\} \subset S^{2n+1}$.

The *complex projective space* $\mathbb{C}P^n$ is defined as the quotient space of the sphere S^{2n+1} by the equivalence relation according to which two points $w, z \in S^{2n+1}$ are equivalent if, and only if, there exists $u \in S^1$ such that $w = u \cdot z$. That is, two points are equivalent if, and only if, they belong to the same orbit. Therefore, for $z \in S^{2n+1}$, its equivalence class is the orbit $\{u \cdot z; u \in S^1\}$ of z.

Each of these equivalence classes is homeomorphic to the circle S^1. The relation we just defined decomposes the sphere S^{2n+1} as a union of pairwise disjoint circles, and each one of them is a point of the complex projective space $\mathbb{C}P^n$.

We denote by $\pi\colon S^{2n+1} \to \mathbb{C}P^n$ the natural projection, which associates to each $z \in S^{2n+1}$ its equivalence class $\pi(z) \in \mathbb{C}P^n$. We endow $\mathbb{C}P^n$ with the quotient topology, according to which a subset $A \subset \mathbb{C}P^n$ is open if, and only if, $\pi^{-1}(A)$ is open in S^{2n+1}. This makes π continuous. The fundamental property of the quotient topology for this particular case can be stated as follows: *For any continuous map $f\colon S^{2n+1} \to Y$ such that $f(u{\cdot}z) = f(z)$ for every $u \in S^1$ and every $z \in S^{2n+1}$, there exists a unique continuous map $\overline{f}\colon \mathbb{C}P^n \to Y$ such that $\overline{f} \circ \pi = f$.*

We say that $\overline{f}$ is obtained from f by *passing to the quotient.*

The natural projection from a space to its quotient space is not necessarily an open map. But, in this case, $\pi\colon S^{2n+1} \to \mathbb{C}P^n$ is open. In order to prove this, take $A \subset S^{2n+1}$ open. For each $u \in S^1$, the set $u \cdot A = \{u \cdot z; z \in A\}$ is open, because $z \mapsto u \cdot z$ is a homeomorphism of S^{2n+1}. Hence, $\pi^{-1}(\pi(A)) = \cup_{u \in S^1} u{\cdot}A$ is open. From the definition of the quotient topology, it follows that $\pi(A)$ is also open in $\mathbb{C}P^n$.

The complex projective space $\mathbb{C}P^n$ is compact, because it is the image of the compact space S^{2n+1} by the continuous map π. It is also a Hausdorff space, as the reader can easily prove.

The decomposition of S^{2n+1} in circles has a structure, called fibration, that occurs in several other geometrical situations. We define this important concept now.

A *locally trivial fibration*, with total space E, *base* B and *typical fiber* F, is a continuous map $\pi\colon E \to B$ with the following property: For every point $x \in B$, there exists a neighborhood $U \ni x$ and a homeomorphism

$$\varphi_U\colon U \times F \to \pi^{-1}(U)$$

such that $\pi \circ \varphi_U = \pi_U$, where $\pi_U\colon U \times F \to U$ is the projection in the first coordinate. That is, the diagram below is commutative.

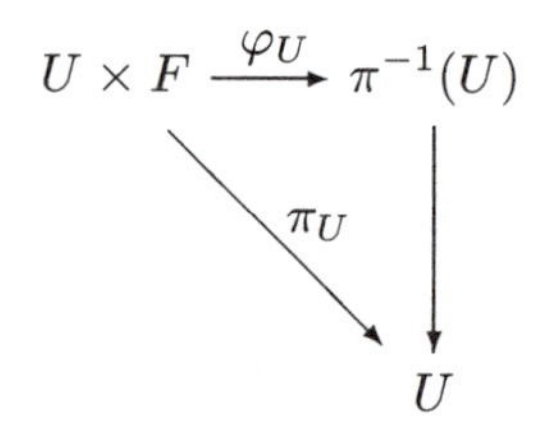

The equality $\pi(\varphi_U(x, y)) = x$ means that, for each $x \in U$, φ_U takes $x \times F$ homeomorphically onto $\pi^{-1}(x)$. Thus, the inverse image $\pi^{-1}(x)$ of each point of B is homeomorphic to the typical fiber F.

Each one of the neighborhoods U above is called a *distinguished neighborhood* and the homeomorphism φ_U is called a *local trivialization*. If we denote by $\psi_U\colon \pi^{-1}(U) \to U \times F$ the inverse homeomorphism of φ_U, we have $\pi_U \circ \psi_U = \pi$ and ψ_U transforms homeomorphically $\pi^{-1}(x)$ onto $x \times F$, for each $x \in U$. In certain situations (as the one we will see now), it may be more convenient to use ψ_U as a local trivialization. Obviously, this does not make any difference.

Every locally trivial fibration is an open map.

The simplest example of a locally trivial fibration is the *product fibration*, where $E = B \times F$ and $\pi = \pi_B \colon B \times F \to B$.

The next stage, in a scale of simplicity, is the *trivial fibration.* Here, $\pi \colon E \to B$ does not appear as the projection in the first factor of a product but there exists a *global trivialization*; that is, a homeomorphism $\varphi = \varphi_B \colon B \times F \to E$ such that $\pi(\varphi(x,y)) = x$ for every $(x,y) \in B \times F$. An example of trivial fibration is given by $\pi \colon \mathbb{R}^n - \{0\} \to S^{n-1}$, $\pi(x) = x/|x|$. In this example, the typical fiber is $\mathbb{R}$ and the global trivialization $\varphi \colon S^{n-1} \times \mathbb{R} \to \mathbb{R}^n - \{0\}$ is $\varphi(x,t) = e^t \cdot x$.

We should remark that the fibration $\pi \colon E \to B$ is called locally trivial because it induces, over each distinguished neighborhood $U \subset B$, the trivial fibration $\pi' \colon \pi^{-1}(U) \to U$, where π' is the restriction of π to $\pi^{-1}(U)$.

The locally trivial fibrations play a relevant role in topology, geometry (differential and algebraic), analysis, physics, and many other areas of pure and applied mathematics.

In order to compute the fundamental group of $\mathbb{C}P^n$, we first show that this space is the base of a locally trivial fibration.

Proposition 3.11. *The quotient map $\pi \colon S^{2n+1} \to \mathbb{C}P^n$ is a locally trivial fibration, with typical fiber S^1.*

Proof. For $j = 1, 2, \ldots, n+1$, the sets $V_j = \{z \in S^{2n+1}; z_j \neq 0\}$ are open in S^{2n+1} therefore the sets $U_j = \pi(V_j)$ are open in $\mathbb{C}P^n$. The sets U_j cover the complex projective space. Moreover, it is easy to see that $V_j = \pi^{-1}(U_j)$. Define the maps $\psi_j \colon V_j \to U_j \times S^1$ by

$$\psi_j(z) = (\pi(z), \frac{z_j}{|z_j|}), \quad j = 1, 2, \ldots, n+1.$$

We have, evidently, $\pi_{U_j} \circ \psi_j = \pi$. In order to prove that each ψ_j is a trivialization, we only need to verify that it is a homeomorphism. The continuity of ψ_j is evident. Define its inverse $\varphi_j \colon U_j \times S^1 \to V_j$ by setting, for each $\pi(z) \in U_j$ and each $u \in S^1$,

$$\varphi_j(\pi(z), u) = \frac{u\overline{z}_j}{|z_j|} \cdot z.$$

The right hand side of the above equation remains the same when we change z by $v \cdot z$, with $v \in S^1$; hence, φ_j is well defined. The fact that φ_j is the inverse of ψ_j can be seen as follows:

$$\varphi_j\psi_j(z) = \varphi_j\left(\pi(z), \frac{z_j}{|z_j|}\right) = \frac{z_j}{|z_j|}\cdot\frac{\overline{z}_j}{|z_j|}\cdot z = z.$$

On the other hand,

$$\psi_j\varphi_j(\pi(z), u) = \psi_j\left(\frac{u\overline{z}_j}{|z_j|}\cdot z\right) = (\pi(z), u).$$

In fact, $v = u\overline{z}_j/|z_j| \in S^1$; hence, $\pi(v\cdot z) = \pi(z)$ and, by defining $w = v\cdot z$ (hence $w_j = vz_j$), an easy computation shows that $w_j/|w_j| = u$.

In order to prove that $\varphi_j\colon U_j \times S^1 \to V_j$ is continuous, we consider $\sigma_j\colon U_j \to V_j$, defined by $\sigma_j(\pi(z)) = (\overline{z}_j/|z_j|)\cdot z$. The right hand side of this equality does not change if we replace z by $u\cdot z$, $u \in S^1$; Hence, σ_j is well defined. Moreover, $\sigma_j \circ \pi\colon V_j \to V_j$ takes z into $(\overline{z}_j/|z_j|)\cdot z$, hence it is continuous. Thus, $\sigma_j\colon U_j \to V_j$ is continuous. Finally, since $\varphi_j(w, u) = u\cdot\sigma_j(w)$, we conclude that φ_j is continuous. □

The following proposition states that a locally trivial fibration has the path lifting property.

Proposition 3.12. *Let $\pi\colon E \to B$ be a locally trivial fibration, $a\colon J \to B$ a path, with $J = [s_0, s_1]$, and $z_0 \in E$ a point such that $\pi(z_0) = a(s_0)$. There exists a path $\widetilde{a}\colon J \to E$ such that $\pi\circ\widetilde{a} = a$ and $\widetilde{a}(s) = z_0$.*

Proof. Suppose, initially, that the image $a(J)$ is contained in a distinguished neighborhood U. We have the local trivialization $\varphi_U\colon U\times F \to \pi^{-1}(U)$, with $z_0 = \varphi_U(a(s_0), y_0)$, $y_0 \in F$. We define the path $\widetilde{a}\colon J \to E$ by $\widetilde{a}(s) = \varphi_U(a(s), y_0)$, for every $s \in J$.

Now we consider the particular case in which $J = J_1 \cup J_2$ is the union of two adjacent compact intervals, with an endpoint s_* in common, in such a way that, by setting $a_1 = a|J_1$ and $a_2 = a|J_2$, the proposition is valid for a_1 and a_2. Then there exists a path $\widetilde{a}_1\colon J_1 \to E$, with $\widetilde{a}_1(s_0) = z_0$ and $\pi\circ\widetilde{a}_1 = a_1$. Let $z_* = \widetilde{a}_1(s_*)$ and $x_* = \pi(z_*) = a(s_*)$. There is a path $\widetilde{a}_2\colon J_2 \to E$, with $\widetilde{a}_2(s_*) = z_*$ and $\pi\circ\widetilde{a}_2 = a_2$. We define the path $\widetilde{a}\colon J \to E$ by $\widetilde{a}|J_1 = \widetilde{a}_1$ and $\widetilde{a}|J_2 = \widetilde{a}_2$.

The general case reduces to the repeated use of the particular case above since, by virtue of the compactness of J, there exists a decomposition $J = J_1 \cup\cdots\cup J_k$ in consecutive intervals such that each image $a(J_i)$ is contained in a distinguished neighborhood. □

The path $\widetilde{a}$ is called a *lifting* of the path a beginning at the point z_0.

Remark. The initial condition $\widetilde{a}(s_0) = z_0$ does not assure us that the lifting $\widetilde{a}$ of the path a is unique, unless the fiber F is totally disconnected.

Later on, we will use the following corollary to Proposition 3.12:

Corollary 3.6. *Let $\pi\colon E \to B$ be a locally trivial fibration. If the base B and the typical fiber F are pathwise connected, the total space E is also pathwise connected.*

In fact, given x, $y \in E$, there exists a path in B connecting $\pi(x)$ to $\pi(y)$. The lifting of this path from the point x connects this point to a point $z \in \pi^{-1}(\pi(y))$. Since the typical fiber is pathwise connected, the same happens with $\pi^{-1}(\pi(y))$; hence, z can be connected to y by a path in this fiber over $\pi(y)$. By composing, we obtain a path in E, connecting x to y.

In the following proposition, we have $z_0 \in E$ and $x_0 = \pi(z_0)$.

Proposition 3.13. *Let $\pi\colon E \to F$ be a locally trivial fibration. If the typical fiber F is pathwise connected, the induced homomorphism $\pi_\#\colon \pi_1(E, z_0) \to \pi_1(B, x_0)$ is surjective.*

Proof. For each $x \in B$, the fiber $\pi^{-1}(x)$ over x is homeomorphic to the typical fiber F; therefore, it is pathwise connected. Let $a\colon I \to B$ be an arbitrary closed path with base x_0. Using Proposition 3.12, we obtain a path $b\colon I \to E$, contained in $\pi^{-1}(x_0)$, connecting $\widetilde{a}(1)$ to the point z_0. We have that $\pi \circ b = e_{x_0}$ is a constant path onto x_0. The path $\widetilde{a}b$ is closed in E, with base z_0, and it satisfies

$$\pi_\#([\widetilde{a}b]) = [\pi \circ (\widetilde{a}b)] = [ae_{x_0}] = [a].$$

Therefore, $\pi_\#$ is surjective. □

Corollary 3.7. *For every $n \geq 1$, the complex projective space $\mathbb{C}P^n$ is simply connected.*

In fact, in the fibration $\pi\colon S^{2n+1} \to \mathbb{C}P^n$, the total space S^{2n+1} is simply connected and the typical fiber S^1 is pathwise connected.

3.5 Exercises

1. An open set of $\mathbb{R}^2$ cannot be homeomorphic to an open set of $\mathbb{R}^n$ if $n > 2$.

2. For any two continuous maps $f,\ g\colon S^1 \to S^1$, the degree of $f \circ g$ is the product of the degrees of f and g.

3. A map $f\colon S^1 \to S^1$ is called *odd* when $f(-x) = -f(x)$ for every $x \in S^1$. Prove that the degree of a continuous odd map is an odd number.

4. Let $T^2 = S^1 \times S^1$ be the torus, and $m,\ n,\ p,\ q$ be integers. Consider the continuous map $f\colon T^2 \to T^2$, defined by

$$f(e^{ix}, e^{iy}) = (e^{i(mx+ny)}, e^{i(px+py)}).$$

Determine the homomorphism $f_{\#}\colon \pi_1(T^2) \to \pi_1(T^2)$, induced by f. Prove that f is a homotopy equivalence if, and only if, it is a homeomorphism.

5. Every retract of a contractible space is contractible. Conclude that the circle S^1 is not the retract of the disk B^2.

6. Let $n > 1$. For every continuous map $f\colon S^n \to \mathbb{R}^2$, there exists $x \in S^n$ such that $f(-x) = f(x)$.

7. Every 3×3 matrix with positive elements has a positive eigenvalue. (Suggestion: Let A be the given matrix. Consider the set P of the elements of S^2 whose three coordinates are non-negative. Find a fixed point of the map $f\colon P \to P$, $f(x) = A(x)/|A(x)|$.)

8. Consider two continuous maps $f,\ g\colon S^2 \to \mathbb{R}$ such that $f(-x) = -f(x)$ and $g(-x) = -g(x)$ for every $x \in S^2$. Prove that there is some point $x_0 \in S^2$ for which we have $f(x_0) = g(x_0) = 0$.

9. Let $\pi\colon E \to B$ be a locally trivial fibration. If the base B and the typical fiber F are connected, then the total space E is also connected.

10. Let $a\colon I \to P^n$ be a closed path such that $d(a(s), a(0)) < \sqrt{2}$ for every $s \in I$. Prove that $[a] = 0$.

11. Let $f,\ g\colon B^2 \to S^2$ be continuous maps, such that $(x, y) \in S^1 \Rightarrow f(x, y) = (x, y, 0)$, $g(x, y) = (-y, x, 0)$. Prove that there exists $(x, y) \in B^2$, with $f(x, y) = \pm g(x, y)$.

Chapter 4

Classical Matrix Groups

In this chapter, we study the classical matrix groups $\mathrm{SO}(n)$, $\mathrm{SU}(n)$, and $\mathrm{Sp}(n)$. We prove some homotopy results related to them, and compute their fundamental groups.

4.1 Rotations in Euclidean Space

Let $\mathrm{SO}(n)$ be the group of rotations of the Euclidean space $\mathbb{R}^n$; that is, the set of all linear transformations $T\colon \mathbb{R}^n \to \mathbb{R}^n$ such that

$$\langle T(x), T(y)\rangle = \langle x, y\rangle$$

for any $x, y \in \mathbb{R}^n$; and $\det T = 1$. The elements of $\mathrm{SO}(n)$ can also be interpreted as real $n \times n$ orthogonal matrices with the determinant equal to 1.

The multiplication (composition) of linear transformations turns $\mathrm{SO}(n)$ into a group, and its natural topology provides a topological group structure. The fact that the determinant of each of its elements is positive implies that $\mathrm{SO}(n)$ is pathwise connected. This is proved right after Proposition. 4.5

Note that the matrices $X \in \mathrm{SO}(n)$ are solutions of the system of quadratic equations $X \cdot X^T = I$. From this, we can prove, using the Implicit Function Theorem, that $\mathrm{SO}(n)$ is a compact surface of dimension $n(n-1)/2$. This is the content of the following proposition.

Proposition 4.1. *The group $\mathrm{SO}(n)$ is a compact surface of dimension $n(n-1)/2$ in the space $\mathbb{R}^{n^2}$ of the $n \times n$ matrices.*

Proof. One form of the Implicit Function theorem says that if $f\colon U \to \mathbb{R}^m$ is a smooth map defined in an open subset U of Euclidean space $\mathbb{R}^n$ and $c \in \mathbb{R}^m$ is a regular value of f (that is, $f'(x)\colon \mathbb{R}^n \to \mathbb{R}^m$ is surjective for every $x \in f^{-1}(c)$), then $f^{-1}(c)$ is a smooth surface of dimension $n-m$ in $\mathbb{R}^n$. (See Bredon (1993), page 84.).

Now let $U \subset \mathbb{R}^n$ be the open set of all invertible $n \times n$ matrices with a positive determinant. Identify the set of symmetric $n \times n$ matrices with $\mathbb{R}^{n(n+1)/2}$. Consider the smooth map $f\colon U \to \mathbb{R}^{n(n+1)/2}$ given by $f(X) = XX^{\mathrm{T}}$. Then $\mathrm{SO}(n) = f^{-1}(I)$, where I is the $n \times n$ identity matrix. To prove that $\mathrm{SO}(n)$ is a smooth surface of dimension $n^2 - n(n+1)/2 = n(n-1)/2$, it suffices to show that the derivative $f'(X)\colon \mathbb{R}^{n^2} \to \mathbb{R}^{n(n+1)/2}$ is surjective for each $X \in \mathrm{SO}(n)$. Clearly, for every $n \times n$ matrix V, we have $f'(X).V = VX^{\mathrm{T}} + XV^{\mathrm{T}}$. Given any symmetric matrix $S \in \mathbb{R}^{n(n+1)/2}$, take $V = SX/2$. Then, since $XX^{\mathrm{T}} = I$ and $S^{\mathrm{T}} = S$, we have

$$f'(X).V = SXX^{\mathrm{T}}/2 + XX^{\mathrm{T}}S/2 = S/2 + S/2 = S,$$

so $f'(X)$ is surjective. The surface $\mathrm{SO}(n)$ is clearly a closed subset of $\mathbb{R}^{n^2}$. It is also bounded since all columns of an orthogonal matrix have length 1. Therefore, $\mathrm{SO}(n)$ is compact. □

For $n = 1$, we have $\mathrm{SO}(1) = \{1\}$, a group with only one element. When $n = 2$, each element of $\mathrm{SO}(2)$ is a 2×2 matrix, whose columns, in the natural order, constitute a positive orthonormal basis of $\mathbb{R}^2$. To determine it, we just have to know the first column, a complex number $z \in S^1$ (since the other must be $i \cdot z$, with $i = \sqrt{-1}$). Thus, $\mathrm{SO}(2)$ is isomorphic to S^1, the multiplicative group of complex numbers of modulus 1. In particular, $\pi_1(\mathrm{SO}(2)) = \mathbb{Z}$.

A geometric interpretation of the above isomorphism can be obtained as follows: A positive rotation of the euclidean plane is completely defined, in a unique way, by a point $z \in S^1$. This point z is a complex number of modulus 1, thus $z = \cos(2\pi\theta) + i\sin(2\pi\theta)$, $\theta \in [0,1)$. The matrix of the rotation is given by

$$R(\theta) = \begin{pmatrix} \cos(2\pi\theta) & -\sin(2\pi\theta) \\ \sin(2\pi\theta) & \cos(2\pi\theta) \end{pmatrix}.$$

For a given point $x = (x_1, x_2) \in \mathbb{R}^2$, its image under the rotation can be obtained by a matrix multiplication $R(\theta) \cdot x$, or by a complex number multiplication zx.

For every $n \geq 3$, we prove in what follows that $\pi_1(\mathrm{SO}(n)) = \mathbb{Z}_2$ is a group of two elements. In this section, we compute the particular cases $\pi_1(\mathrm{SO}(3)) = \mathbb{Z}_2$ and $\pi_1(\mathrm{SO}(4)) = \mathbb{Z}_2$. This is done by proving that $\mathrm{SO}(3)$

is homeomorphic to the real projective space P^3. This is an exceptional fact. There is no relation between P^n and SO(n) when $n > 3$. (Note the coincidence of the dimensions: $n(n-1)/2 = 3$ if, and only if, $n = 3$.)

In order to relate SO(3) with P^3, we use the algebra of quaternions, created by the Irish mathematician Sir William Hamilton.

The set of quaternions is simply the Euclidean space $\mathbb{R}^4$ in which we introduce a multiplication with interesting properties.

Each quaternion (element of $\mathbb{R}^4$) will be represented in the form

$$w = t + xi + yj + zk$$

instead of $w = (t, x, y, z)$. The basic vectors 1, i, j, k are called the *unit* quaternions. $1 = 1+0{\cdot}i+0{\cdot}j+0{\cdot}k$ is the real unity; the others are imaginary units. The operations of vector space are the usual ones in $\mathbb{R}^4$.

In the space of the quaternions $\mathbb{R}^4$, we single out two special subspaces: $\mathbb{R}$ and $\mathbb{R}^3$. $\mathbb{R}$ is the set of real quaternions $t + 0{\cdot}i + 0{\cdot}j + 0{\cdot}k$ and $\mathbb{R}^3$ is the set of pure imaginary quaternions $xi + yj + z{\cdot}k$. We also say, in accordance with the traditional Hamiltonian vector calculus, that $\mathbb{R}$ is the set of scalars and $\mathbb{R}^3$ is the set of vectors. With respect to the standard inner product of $\mathbb{R}^4$ (which we will always adopt), $\mathbb{R}^3$ is the orthogonal complement of $\mathbb{R}$.

Quaternion multiplication is defined, by bilinearity, when we define the products of the unit elements. This is shown in the following table:

	1	i	j	k
1	1	i	j	k
i	i	-1	k	$-j$
j	j	$-k$	-1	i
k	k	j	$-i$	-1

The above multiplication has the properties of distributivity (seen by brute force) and associativity (tedious verification). Nevertheless, commutativity does not hold, as shown by the table itself.

On the other hand, every non-null quaternion w has a multiplicative inverse w^{-1}. To prove this fact, we introduce the *conjugate* $\overline{w}$ of a quaternion w:

$$\overline{w} = t - xi - yj - zk \quad \text{if} \quad w = t + xi + yj + zk.$$

We have $w{\cdot}\overline{w} = \overline{w}{\cdot}w = |w|^2$ where $|w|^2 = t^2 + x^2 + y^2 + z^2$. Therefore, if $w \neq 0$, the definition

$$w^{-1} = \frac{\overline{w}}{|w|^2}$$

yields $w{\cdot}w^{-1} = w^{-1}{\cdot}w = 1$.

The modulus of a quaternion behaves well with respect to the operation of quaternion multiplication: $|w \cdot w'| = |w| \cdot |w'|$.

From this, it follows that the sphere $S^3 \subset \mathbb{R}^4$, the set of quaternions of modulus 1, is a group with respect to quaternion multiplication. The multiplication $\mathbb{R}^4 \times \mathbb{R}^4 \to \mathbb{R}^4$, being bilinear, is continuous. Hence, S^3 is a topological group. (When $|w| = 1$, we have $w^{-1} = \overline{w}$.)

Lemma 4.1. *If the quaternion w commutes with every pure imaginary quaternion then w is real. If, moreover, $w \in S^3$, then $w = \pm 1$.*

Proof. If $w = a + bi + cj + dk$ then $iw = -b + ai - dj + ck$, and $wi = -b + ai + dj - ck$. From $wi = iw$, we conclude that $c = d = 0$; that is, $w = a + bi$. Hence, $wj = aj + bk$ and $jw = aj - bk$. From this, we have (using $wj = jw$) that $b = 0$; therefore, $w = a$ is real. □

Proposition 4.2. *There exists a continuous and surjective homomorphism $\varphi\colon S^3 \to \mathrm{SO}(3)$, whose kernel is $\{1, -1\}$.*

Proof. To each $u \in S^3$ we associate the linear transformation $\varphi_u \colon \mathbb{R}^3 \to \mathbb{R}^3$, defined by $\varphi_u(w) = u \cdot w \cdot u^{-1}$. First we prove that φ_u is well defined. Considered initially as defined in $\mathbb{R}^4$, φ_u is evidently linear and, since $|u \cdot w \cdot u^{-1}| = |w|$, $\varphi_u \colon \mathbb{R}^4 \to \mathbb{R}^4$, it is orthogonal. Moreover, since $\varphi_u(1) = 1$, the subspace $\mathbb{R}$ of the reals is invariant by φ. Therefore, its orthogonal complement $\mathbb{R}^3$, the set of pure imaginaries, is also invariant by φ. In other words, when $w = xi + yj + zk$ is a pure imaginary, the same holds for $u{\cdot}w{\cdot}u^{-1}$. Thus, the orthogonal linear transformation $\varphi_u \colon \mathbb{R}^3 \to \mathbb{R}^3$ is well defined.

The columns of the matrix of φ_u are the vectors $u{\cdot}i{\cdot}u^{-1}$, $u{\cdot}j{\cdot}u^{-1}$ and $u{\cdot}k{\cdot}u^{-1}$, which depend continuously on $u \in S^3$. We have $\det(\varphi_u) = \pm 1$ for every $u \in S^3$. Since S^3 is connected and, for $u = 1$, we have $\det(\varphi_u) = 1$, it follows that $\det(\varphi_u) = 1$ for every $u \in S^3$. Therefore, $\varphi_u \in \mathrm{SO}(3)$ for every $u \in S^3$, which gives us a continuous map

$$\varphi\colon S^3 \to \mathrm{SO}(3), \quad u \mapsto \varphi_u = \varphi(u).$$

Evidently, $\varphi_{uv} = \varphi_u \circ \varphi_v$, so φ is a group homomorphism. The kernel of φ is the set of all quaternions $u \in S^3$ such that $u{\cdot}w{\cdot}u^{-1} = w$; that is, $u{\cdot}w = w{\cdot}u$, for every $w \in \mathbb{R}^3$. By Lemma 4.1, we conclude that the kernel of φ contains only the quaternions 1 and -1. In other words, $\varphi(x) = \varphi(y) \Leftrightarrow y = \pm x$. In particular, φ is locally injective.

In order to conclude the proof, we need to prove that φ is surjective. Since S^3 is compact and $\mathrm{SO}(3)$ is connected, it is enough to show that φ is an open map. (In fact, it follows from this that $\varphi(S^3)$ is a closed and open subset of SO(3).) With this in mind, we appeal to differential calculus. We start by observing that φ is a map of class C^∞. (For each $u \in S^3$, the elements of the matrix of φ_u are infinitely differentiable functions of u.) Since φ is a group homomorphism, its rank is constant. From the rank theorem, since φ is locally injective, its rank is maximum, that is, it is equal to 3. In particular, φ is a local diffeomorphism and therefore an open map. □

For information about the rank theorem, used in the above proof, the reader should consult Dieudonné (1960), page 273. A simpler proof can be found in Lima (1999), page 300 (in Portuguese).

For a purely algebraic proof that φ is surjective, see Exercise 6 at the end of this chapter.

Remark. Let $f\colon M \to N$ be a smooth map between surfaces. At every point $x \in M$, the derivative of f is a linear map $f'(x)\colon T_xM \to T_yN$, $y = f(x)$. The rank of this linear map, i.e., the dimension of its image space, is called the *rank of f at the point x*. When the surfaces have a group structure, with smooth multiplication and inversion, and the map between them is a smooth homomorphism $\varphi\colon G \to H$, then the rank of φ is the same at all points $x \in G$. To see this, let $x_1, x_2 \in G$ be any two points, with $y_1 = \varphi(x_1)$ and $y_2 = \varphi(x_2)$. The left translation by $x_2x_1^{-1}$ in G and by $y_2y_1^{-1}$ in H are diffeomorphisms $\lambda\colon G \to G$, $\mu\colon H \to H$ such that $\varphi \circ \lambda = \mu \circ \varphi$; hence, $\varphi'(x_1).\lambda'(x_1) = \mu'(y_1).\varphi'(x_1)$. Since $\lambda'(x_1)\colon T_{x_1}G \to T_{x_2}G$ and $\mu'(x_1)\colon T_{y_1}H \to T_{y_2}H$ are isomorphisms, it follows that the linear maps $\varphi'(x_1)\colon T_{x_1}G \to T_{y_1}H$ and $\varphi'(x_2)\colon T_{x_2}G \to T_{y_2}H$ have the same rank. In the case of the above lemma, the homomorphism $\varphi\colon S^3 \to SO(3)$ is locally injective, since $\varphi(x) = \varphi(x') \Rightarrow x' = -x$. By the rank theorem, a map of constant rank can only be locally injective when its derivative is injective at each point. Since S^3 and $SO(3)$ have the same dimension, this means that $\varphi'(x)$ is an isomorphism at each point $x \in S^3$ so, by the Inverse Function theorem, φ is a local diffeomorphism; hence, it an open map.

Corollary 4.1. *The group* SO(3) *of rotations of Euclidean space* $\mathbb{R}^3$ *is homeomorphic to the projective space* P^3.

In fact, since the kernel of φ is $\{+1, -1\}$, we have $\varphi(w) = \varphi(w') \Leftrightarrow w' = \pm w$. By passing to the quotient, we obtain a continuous bijection $\overline{\varphi}\colon P^3 \to \mathrm{SO}(3)$. Since P^3 is compact and SO(3) is Hausdorff, $\overline{\varphi}$ is a homeomorphism.

It follows from Corollary 4.1 that $\pi_1(\mathrm{SO}(3)) \approx \pi_1(P^3) \approx \mathbb{Z}_2$.

Remark. Define a map $h\colon S^3 \to S^2$ by setting, for each $u \in S^3$, $h(u) = u^{-1}\cdot i\cdot u = \varphi_u(i)$. The map h is continuous, surjective, and $h(u) = h(v) \Leftrightarrow u\cdot v^{-1}$ commutes with i, which means that $w = u\cdot v^{-1} = a + bi$ is a complex number. The result from this is the equivalence relation induced in S^3 by h is the same that defines $\mathbb{C}P^1$ as a quotient space. By passing to the quotient, we obtain a homeomorphism $\overline{h}\colon \mathbb{C}P^1 \to S^2$. Another result from this is h is a locally trivial fibration with typical fiber S^1.

The map h, introduced in the above remark, is known as *Hopf fibration*, in honor of the topologist Heinz Hopf, who introduced it. It is a classical object and constitutes an outstanding mark in the history of topology.

In order to exhibit explicitly a closed path whose homotopy class is the non-null element of $\pi_1(\mathrm{SO}(3))$, it is enough to consider a path in S^3, such as $\widetilde{a}\colon I \to S^3$ defined by $\widetilde{a}(s) = \cos \pi s + \sin \pi s\cdot k$, with antipodal endpoints $\widetilde{a}(0) = 1$, $\widetilde{a}(1) = -1$. Then $a = \varphi \circ \widetilde{a}\colon I \to \mathrm{SO}(3)$ is a closed path non homotopic to a constant. For each $s \in I$, the columns of the matrix of the linear transformation $a(s) = \varphi_{\widetilde{a}(s)}$ are the images of the unit quaternions $i, j, k \in \mathbb{R}^3$, which are

$$\begin{aligned} a(s)(i) &= (\cos \pi s + \sin \pi s\cdot k)\cdot i\cdot(\cos \pi s - \sin \pi s\cdot k) = \\ &= (\cos^2 \pi s - \sin^2 \pi s)\cdot i + 2 \sin \pi s \cos \pi s\cdot j = \\ &= \cos 2\pi s\cdot i + \sin 2\pi s\cdot j, \\ a(s)(j) &= -sen 2\pi s\cdot i + \cos 2\pi s\cdot j \ \text{ and} \\ a(s)(k) &= k. \end{aligned}$$

Therefore, the generator of $\pi_1(\mathrm{SO}(3))$ is the homotopy class of the closed path $a\colon I \to \mathrm{SO}(3)$ such that, for each $s \in I$, the transformation $a(s)\colon \mathbb{R}^3 \to \mathbb{R}^3$ has the matrix

$$\begin{pmatrix} \cos 2\pi s & -\sin 2\pi s & 0 \\ \sin 2\pi s & \cos 2\pi s & 0 \\ 0 & 0 & 1 \end{pmatrix}.$$

Remark. An intuitive explanation for the fact that aa is homotopic to a constant (that is, $a \cong a^{-1}$), can be given as follows: For each $s \in I$, the columns of the matrix $a(s)$ form a positive orthonormal trihedron whose two first vectors are on the horizontal plane and the third is the vector $k = (0, 0, 1)$. (See Figure 4.1.)
When s varies in I, the two first vectors of $a(s)$ describe the equatorial circle and the third remains fixed. The homotopy between a and a^{-1} consists of turning the equatorial plane by an angle of 180° around an axis E. For each instant t we have a path analogous to a but made up of trihedra whose first

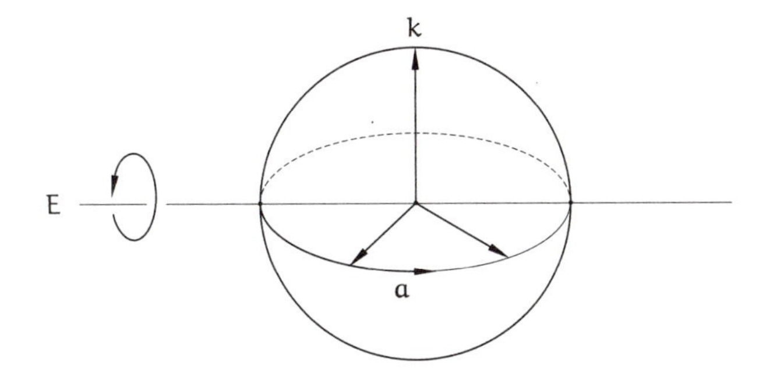

Figure 4.1.

two vectors are on the plane that passes through the axis E and makes an angle of πt radians with the horizontal plane. At the end of the homotopy we have the path a^{-1}, which consists of a trihedron whose two first vectors describe the horizontal equatorial circle in the opposed sense of the initial one and the third vector remains fixed, equal to $-k$.

The computation of the fundamental group of SO(4) is based in the following proposition.

Proposition 4.3. *The topological space* SO(4) *is homeomorphic to the Cartesian product of* SO(3) *and the sphere* S^3.

Proof. Let $h\colon \mathrm{SO}(4) \to \mathrm{SO}(3) \times S^3$ be the continuous map that associates to each orthogonal operator $T\colon \mathbb{R}^4 \to \mathbb{R}^4$ the pair $h(T) = (T', w)$, where $w = T(1)$ is the quaternion of modulus 1, image of the quaternion 1 by T, and $T'\colon \mathbb{R}^3 \to \mathbb{R}^3$ is defined by $T'(v) = T(v)\cdot w^{-1}$ (quaternion multiplication) for every $v \in \mathbb{R}^3$. The map T' is, by its definition, an operator of $\mathbb{R}^4$ but, since $T'(1) = 1$, its restriction to $\mathbb{R}^3$ takes the pure imaginary v into another pure imaginary $T'(v)$. It is easy to prove that h is bijective; hence, it is a homeomorphism. □

Corollary 4.2. *The fundamental group of* SO(4) *is* $\mathbb{Z}_2$.

In fact,

$$\pi_1(\mathrm{SO}(4)) = \pi_1(\mathrm{SO}(3)) \times \pi_1(S^3) = \pi_1(\mathrm{SO}(3)) = \mathbb{Z}_2,$$

because the sphere S^3 is simply connected.

From the above homeomorphism and the identification of a generator of $\pi_1(\mathrm{SO}(3))$, it follows that the fundamental group of SO(4) is generated by the homotopy class of the path $b\colon I \to \mathrm{SO}(4)$ where, for every $s \in I$, the operator $b(s)\colon \mathbb{R}^4 \to \mathbb{R}^4$ has the matrix

$$\begin{pmatrix} 1 & 0 & 0 & 0 \\ 0 & \cos 2\pi s & -\sin 2\pi s & 0 \\ 0 & \sin 2\pi s & \cos 2\pi s & 0 \\ 0 & 0 & 0 & 1 \end{pmatrix}.$$

Using the fundamental group of SO(3), we provide below a self-contained proof of the theorem of Poincaré, according to which every continuous tangent vector field to the sphere S^2 has a singularity; that is, a point $x_0 \in S^2$ where the vector field is null. We remind the reader that $v(x)$ tangent to S^2 at the point x means that $v(x)$ is orthogonal to x.

Proposition 4.4. **(Poincaré)** *There is no continuous map* $v\colon S^2 \to \mathbb{R}^3$ *such that* $v(x) \neq 0$ *and* $\langle x, v(x)\rangle = 0$ *for every* $x \in S^2$.

Proof. Suppose, by contradiction, that v exists. By replacing $v(x)$ with $v(x)/|v(x)|$, we obtain a continuous map $v\colon S^2 \to S^2$ such that $\langle x, v(x)\rangle = 0$ for every $x \in S^2$. Then, taking $w(x) = x \times v(x)$ (vector product) we have, for each $x \in S^2$, an orthogonal matrix $M(x) = [x, v(x), w(x)]$, whose columns $x, v(x), w(x)$ depend continuously on x. We have $M(x) \in \mathrm{SO}(3)$. Let's consider $\mathrm{SO}(2) \subset \mathrm{SO}(3)$ using the identification

$$\begin{pmatrix} a & b \\ c & d \end{pmatrix} \longleftrightarrow \begin{pmatrix} 1 & 0 & 0 \\ 0 & a & b \\ 0 & c & d \end{pmatrix}.$$

Now we define a continuous map

$$h\colon S^2 \times \mathrm{SO}(2) \to \mathrm{SO}(3)$$

by $h(x, L) = M(x){\cdot}L$. The map h has an inverse $T \mapsto (x, M(x)^{-1}{\cdot}T)$, where, for every $T \in \mathrm{SO}(3)$, we define $x = T(e_1) =$ first column of T. Then h is a homeomorphism. But, since $\pi_1(S^2 \times \mathrm{SO}(2)) = \mathbb{Z}$ and $\pi_1(\mathrm{SO}(3)) = \mathbb{Z}_2$, these spaces cannot be homeomorphic. This contradiction completes the proof. □

4.2 The Groups SU(n) and Sp(n)

The results of this section will not be used in the chapters that follow.

Based on the cases where that we have studied in the previous section $n \leq 4$, we now compute the fundamental group of SO(n) for every value of n. The method used also allows us to compute the fundamental group of other classical matrix groups, namely: SU(n), U(n), and Sp(n).

The main tool to compute $\pi_1(\mathrm{SO}(n))$ is the map

$$\pi\colon \mathrm{SO}(n) \to S^{n-1},$$

which associates to each linear transformation $T \in \mathrm{SO}(n)$ the unit vector $\pi(T) = T(e_1)$, image by T of the first vector e_1 of the standard basis of $\mathbb{R}^n$. Identifying T with its matrix relative to this standard basis, $\pi(T)$ is simply the first column of the matrix T.

The most important property of the map π is the following proposition.

Proposition 4.5. *The map π is a locally trivial fibration, with typical fiber* $\mathrm{SO}(n-1)$.

Proof. Consider initially the open set $V \subset S^{n-1}$, which consists of the unit vectors $x = (x_1, \ldots, x_n)$ such that $x_1 > 0$. This means that the matrix $[x, e_2, \ldots, e_n]$, whose first column is x, has a positive determinant. The classical orthonormalization process of Gram-Schmidt, applied to the column vectors of this matrix, furnishes an orthogonal matrix with positive determinant, denoted by $\sigma(x)$, which depends continuously on the vector x. In this way, we have defined a continuous map $\sigma\colon V \to \mathrm{SO}(n)$. Since x is a unit vector, the Gram-Schmidt process does not change it; therefore, the first column of the matrix $\sigma(x)$ is x. Hence, we have

$$\sigma\colon V \to \pi^{-1}(V)$$

and σ is a *local section* of π; that is, we have $\pi \circ \sigma = id_V$. From σ, we define a local trivialization

$$\varphi_V\colon V \times \mathrm{SO}(n-1) \to \pi^{-1}(V)$$

by

$$\varphi_V(x, M) = \sigma(x){\cdot}M.$$

Here, we are considering $\mathrm{SO}(n-1) \subset \mathrm{SO}(n)$, where each linear transformation $M \in \mathrm{SO}(n-1)$ operates in $\mathbb{R}^n$ by leaving the vector e_1 fixed. In terms of matrices, this corresponds to identifying each matrix $M \in \mathrm{SO}(n-1)$ with the $n \times n$ matrix obtained from it by inserting the first row $(1, 0, \ldots, 0)$ and the first column also equal to $(1, 0, \ldots, 0)$.

In the definition of φ_V, $\sigma(x) \cdot M$ is a product of matrices. As we can easily prove, the first columns of $\sigma(x)$ and M being respectively x and e_1, the first column of $\sigma(x){\cdot}M$ is x. In other words, we have $\pi(\varphi_V(x, M)) = x$; therefore, φ_V is, indeed, a local trivialization, whose inverse homeomorphism is $\psi_V\colon \pi^{-1}(V) \to V \times \mathrm{SO}(n-1)$, defined by $\psi_V(T) = (x, \sigma(x)^{-1}{\cdot}T)$, where $x = \pi(T)$.

The open set V is a neighborhood of e_1 in S^{n-1}. In order to obtain a local trivializations over neighborhoods of the other points, we take for each $y \in S^{n-1}$, a transformation $T \in \mathrm{SO}(n)$ such that $T(e_1) = y$. Then $W = T(V)$ is an open neighborhood of y. We define the local trivialization

$$\varphi_W : W \times \mathrm{SO}(n-1) \to \pi^{-1}(W)$$

by setting, for every $w \in W$ and every $M \in \mathrm{SO}(n-1)$, $\varphi_W(w, M) = T\sigma T^{-1}(w) \cdot M$. □

The fibration π allows us to conclude, by induction, starting with $\mathrm{SO}(1) = \{1\}$, that $\mathrm{SO}(n)$ is pathwise connected. (See Corollary 3.6.)

Remark. Given S, $T \in \mathrm{SO}(n)$, we have $\pi(S) = \pi(T)$ if, and only if, $S^{-1}T(e_1) = e_1$, that is, $T^{-1}S \in \mathrm{SO}(n-1)$. This means that the fibers $\pi^{-1}(x)$, $x \in S^{n-1}$, are the cosets $T \cdot \mathrm{SO}(n-1)$, relative to the subgroup $\mathrm{SO}(n-1)$. By passing to the quotient, we obtain a continuous bijection $\overline{\pi} \colon \mathrm{SO}(n)/\mathrm{SO}(n-1) \to S^{n-1}$, which is a homeomorphism, because π is open. Thus, the sphere S^{n-1} can be considered as the homogeneous space $\mathrm{SO}(n)/\mathrm{SO}(n-1)$. This is not a quotient group because $\mathrm{SO}(n-1)$ is not a normal subgroup of $\mathrm{SO}(n)$.

In order to establish an analogous fibration for complex matrices, we recall that, in the vector space $\mathbb{C}^n$, whose elements are lists $z = (z_1, \ldots, z_n)$, $w = (w_1, \ldots, w_n)$, of n complex numbers, the *hermitian inner product* is defined by

$$\langle z, w \rangle = z_1\overline{w}_1 + \cdots + z_n\overline{w}_n.$$

The *unitary group* $\mathrm{U}(n)$ is formed by the linear transformations $T \colon \mathbb{C}^n \to \mathbb{C}^n$ which preserve this inner product; that is, which fulfill the condition $\langle T(z), T(w) \rangle = \langle z, w \rangle$ for any z, $w \in \mathbb{C}^n$. Identifying T with its matrix relative to the standard basis of $\mathbb{C}^n$, we can consider $\mathrm{U}(n)$ as the set of complex $n \times n$ matrices whose columns (and rows) have length 1 and are pairwise orthogonal (unitary matrices).

If we use the notation T^* for the transpose matrix of the conjugate of T, we can easily prove that $T \in \mathrm{U}(n) \Leftrightarrow T\,T^* = T^*T = I$ (where I denotes the identity matrix of order n). By taking determinants of both sides, we have

$$T \in \mathrm{U}(n) \Rightarrow \det(T) \cdot \overline{\det(T)} = \det(T\,T^*) = \det I = 1;$$

hence, $\det(T) \in S^1$ for every $T \in \mathrm{U}(n)$.

The description by matrices shows that $\mathrm{U}(n)$ is a bounded and closed subset of $\mathbb{C}^{n^2}$ (or of $\mathbb{R}^{2n^2}$); hence, it is compact.

The *special unitary group* $\mathrm{SU}(n)$ is the subgroup of $\mathrm{U}(n)$ that consists of the unitary matrices which have determinant equal to 1. As a closed subset of $\mathrm{U}(n)$, the group $\mathrm{SU}(n)$ is compact.

We define, as before, a continuous map

$$\pi\colon\ \mathrm{SU}(n) \to S^{2n-1},$$

by $\pi(T) = T(e_1)$. That is, $\pi(T)$ is the first column of the matrix of T relative to the standard basis of $\mathbb{C}^n$.

In order to prove that π is a locally trivial fibration, we consider the open set $V \subset S^{2n-1}$ formed by the unit vectors $z = (z_1, \ldots, z_n)$ such that $z_1 \neq 0$, hence $\{z, e_2, \ldots, e_n\} \subset \mathbb{C}^n$ is a basis. We apply the Gram-Schmidt process to this set, to obtain an orthonormal basis $\{z, v_2, \ldots, v_n\}$. Let Δ be the determinant of this system of vectors. Then $z, v_2/\Delta, v_3, \ldots, v_n$ are columns of a unitary matrix $\sigma(z)$, with determinant 1, which has first column z and depends continuously on z. This defines, therefore, a continuous map

$$\sigma\colon V \to \pi^{-1}(V),$$

with $\pi \circ \sigma = id_V$; hence, σ is a section of π over V. From this, we define the local trivialization

$$\varphi_V\colon V \times \mathrm{SU}(n-1) \to \pi^{-1}(V)$$

by $\varphi_V(z, M) = \sigma(z) \cdot M$.

Here, we are considering $\mathrm{SU}(n-1) \subset \mathrm{SU}(n)$, identifying each matrix $M \in \mathrm{SU}(n-1)$ with the matrix of $\mathrm{SU}(n)$ obtained by inserting in M a first row and a first column, both equal to $(1, 0, \ldots, 0)$.

The continuous map

$$\psi_V\colon \pi^{-1}(V) \to V \times \mathrm{SU}(n-1),$$

defined by

$$\psi_V(T) = (z, \sigma(z)^{-1} \cdot T),$$

where $z = T(e_1)$, is the inverse of φ_V; hence, φ_V is a homeomorphism. Obviously $\pi(\varphi_V(z, M)) = z$; therefore, φ_V is a local trivialization. As in the case of $\mathrm{SO}(n)$, for each point $y \in S^{2n-1}$ we consider a transformation $T \in \mathrm{SU}(n)$ such that $T(e_1) = y$. Then $W = T(V)$ is a neighborhood of y in S^{2n-1}, and we define the local trivialization

$$\varphi_W\colon W \times \mathrm{SU}(n-1) \to \pi^{-1}(W)$$

by $\varphi_W(w, M) = T\sigma T^{-1}(w) \cdot M$. The verification of the details is easy, so we may state that the map $\sigma\colon\ \mathrm{SU}(n) \to S^{2n-1}$ is a locally trivial fibration.

By Corollary 3.6, we conclude, using induction, that $\mathrm{SU}(n)$ is pathwise connected. In fact, $\mathrm{SU}(1) = \{1\}$ is pathwise connected. The base and typical fiber (equal to $\mathrm{SU}(1)$) of the locally trivial fibration $\pi\colon \mathrm{SU}(2) \to S^3$ are pathwise connected. (In fact, π is a homeomorphism between $\mathrm{SU}(2)$ and S^3 since the fiber has only one element.) Now, the fibration $\pi\colon \mathrm{SU}(3) \to S^5$ has base and typical fiber (equal to $\mathrm{SU}(2)$) connected; hence, $\mathrm{SU}(3)$ is pathwise connected. And so on.

As to the group $\mathrm{U}(n)$ of the unitary $n \times n$ matrices, we remark that we have a homeomorphism

$$\mathrm{U}(n) \approx S^1 \times \mathrm{SU}(n).$$

In particular, $\mathrm{U}(n)$ is pathwise connected.

The above homeomorphism is the map $f\colon S^1 \times \mathrm{SU}(n) \to \mathrm{U}(n)$, defined by $f(u, T) = R$, where R is the matrix obtained from T multiplying the first column by u. The inverse of f is the continuous map $g\colon \mathrm{U}(n) \to S^1 \times \mathrm{SU}(n)$, given by $g(R) = (u, T)$, where $u = \det. R$ and T is obtained from R by dividing the first column by u.

There exists a fibration $\pi\colon \mathrm{U}(n) \to S^{2n-1}$, defined in a similar way as the previous ones, but it is not necessary to consider it in order to compute the fundamental group of $\mathrm{U}(n)$ because $\pi_1(\mathrm{U}(n)) = \pi_1(\mathrm{SU}(n)) \times \mathbb{Z}$, by virtue of the above homeomorphism.

The computation of the fundamental group of $\mathrm{SU}(n)$ is based on the following proposition.

Proposition 4.6. *Let $\pi\colon E \to B$ be a locally trivial fibration. If the base B is simply connected and the typical fiber F is pathwise connected, then the fundamental group of E is isomorphic to a quotient group of the fundamental group of F.*

Corollary 4.3. *For every n, $\mathrm{SU}(n)$ is simply connected and $\pi_1(\mathrm{U}(n)) \approx \mathbb{Z}$.*

In fact, $\mathrm{SU}(1) = \{1\}$ and $\mathrm{SU}(2) = S^3$ are simply connected. By its turn, the fibration $\mathrm{SU}(3) \to S^5$, with a simply connected base, shows that $\pi_1(\mathrm{SU}(3))$ is a quotient group of $\pi_1(\mathrm{SU}(2))$, hence it is equal to zero. And so on. Since $\mathrm{U}(n)$ is homeomorphic to the product of S^1 by the simply connected space $\mathrm{SU}(n)$, it follows that $\pi_1(\mathrm{U}(n)) \approx \mathbb{Z}$.

The computation of $\pi_1(\mathrm{SO}(n))$ is based on Proposition 4.7, where the hypothesis contains the statement $\pi_2(B) = 0$. The assertion $\pi_2(B) = 0$ means that every continuous map $f\colon S^2 \to B$ is homotopic to a constant.

Taking into account that the sphere S^2 is homeomorphic to the quotient space of the square $I \times I$ by the equivalence relation that identifies all of the boundary $\partial(I \times I)$ to a single point, we can express the condition $\pi_2(B) = 0$ by saying that, for any given continuous map $g\colon I \times I \to B$ with $g|\partial(I \times I)$

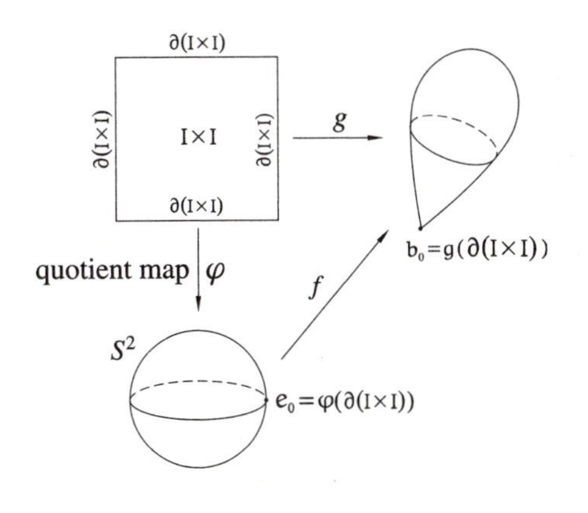

Figure 4.2.

constant, there exists an homotopy $H\colon (I \times I) \times I \to B$, between g and a constant map, such that, for each $t \in I$, $H_t(\partial(I \times I)) = b_t$ reduces to a single point (see Figure 4.2).

In this definition, the base point $H_t(\partial(I \times I) = b_t$ can move during the homotopy. Nevertheless, as we prove now, given the continuous map $g\colon (I \times I, \partial(I \times I)) \to (B, b_0)$ satisfying the above conditions, it is possible to modify the homotopy H so that the base point b_t remains fixed, equal to b_0, for every value of t.

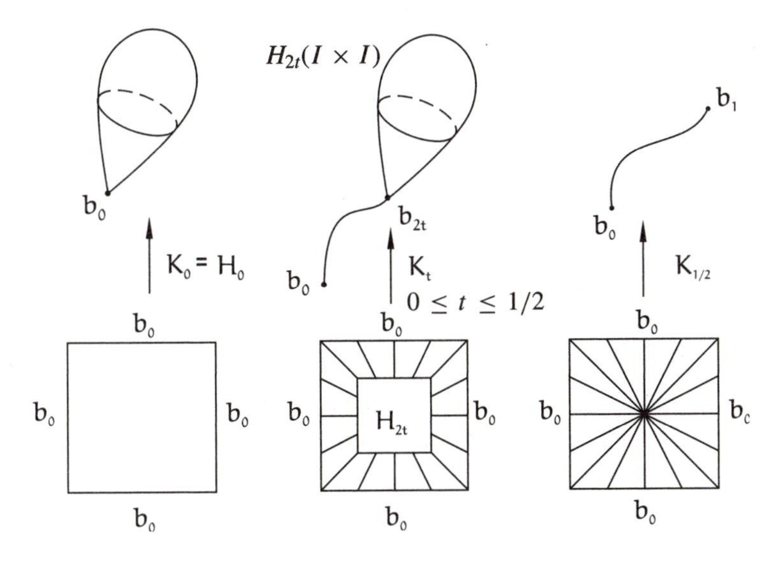

Figure 4.3.

The new homotopy K_t, between g and the constant map $I \times I \to b_0$, with $K_t(\partial(I \times I)) = b_0$ for every t, is illustrated in Figure 4.3. For $0 \leq t \leq 1/2$, the continuous map $K_t \colon I \times I \to B$ takes each of the radial segments into the path that goes from b_0 to b_{2t} and, in the square of face $1-2t$, concentric with $I \times I$, K_t reproduces, in scale, the map H_{2t}. Thus, $K_{1/2}$ transforms all of the square $I \times I$ in the image of the path described by the points $b_t = H_t(\partial(I \times I))$, in such a way that on each ray that starts in the center of $I \times I$, $K_{1/2}$, it moves along the same path.

For $1/2 \leq t \leq 1$, K_t simply contracts the path that goes from b_0 to $b_{2(1-t)}$, back to b_0, maintaining b_0 as fixed.

Proposition 4.7. *Let $\pi \colon E \to B$ be a locally trivial fibration such that the typical fiber F is pathwise connected, the base B is simply connected, and, moreover, $\pi_2(B) = 0$. Then the fundamental group of E is isomorphic to the fundamental group of F.*

Remark. The hypothesis $\pi_2(B) = 0$ in Proposition 4.7 is necessary, as evidenced by the Hopf fibration $S^3 \to S^2$, whose base is simply connected, and the fiber S^1 is pathwise connected but $\pi_1(S^3) \neq \pi_1(S^1)$.

Corollary 4.4. *For $n \geq 3$, we have $\pi_1(\mathrm{SO}(n)) = \mathbb{Z}_2$.*

In fact, we know that $\pi_1(\mathrm{SO}(3)) = \mathbb{Z}_2$ and that $\mathrm{SO}(n)$ is a locally trivial fiber space with base S^{n-1} and typical fiber $\mathrm{SO}(n-1)$. Moreover, for $n \geq 3$, the sphere S^{n-1} is simply connected. We just have to prove that $\pi_2(S^n) = 0$ for $n \geq 3$. For this, we observe that every continuous map $f \colon S^2 \to S^n$ is homotopic to a differentiable map, whose image has measure zero (Sard's theorem) when $n > 2$. In particular, a differentiable map $S^2 \to S^n$ is never surjective if $n > 2$; hence, it is homotopic to a constant.

In order to complete the computation of $\pi_1(\mathrm{SO}(n))$, $\pi_1(\mathrm{SU}(n))$ and $\pi_1(\mathrm{U}(n))$, it remains to prove Propositions 4.6 and 4.7.

The proof of Proposition 4.6 uses the lemma below. We say that a homotopy $H \colon X \times I \to Y$ *starts* with $f \colon X \to Y$ when $H(x, 0) = f(x)$ for every $x \in X$. For an arbitrary fibration $\pi \colon E \to B$, we say that $\widetilde{H} \colon X \times I \to E$ is a *lifting* of the homotopy $H \colon X \times I \to B$ when $\pi \circ \widetilde{H} = H$.

Lemma 4.2. (Homotopy lifting for paths) *Let $\pi \colon E \to B$ be a locally trivial fibration. Given a path $\widetilde{a} \colon J \to E$, every homotopy $H \colon J \times I \to B$ that starts with $a = \pi \circ \widetilde{a}$ has a lifting $\widetilde{H} \colon J \times I \to E$ that starts with $\widetilde{a}$.*

In general, Lemma 4.2 is used in a seemingly stronger version, where we impose to the lifting $\widetilde{H}$ not only that it starts with $\widetilde{a}$ but also that it satisfies the additional conditions that are specified in the next lemma.

Let $J = [s_0, s_1]$, so $\partial J = \{s_0, s_1\}$. In the lemma below, X represents one of the sets $(J \times 0) \cup (\partial J \times I)$ or $(J \times 0) \cup (s_1 \times I)$, that is, the base and the vertical sides of the rectangle $J \times I$ or the base and the vertical left side of the rectangle.

Lemma 4.3. *Let $\pi \colon E \to B$ be a locally trivial fibration and $\widetilde{f} \colon X \to E$ be a continuous map. Every homotopy $H \colon J \times I \to B$ that coincides with $f = \pi \circ \widetilde{f}$ in X has a lifting $\widetilde{H} \colon J \times I \to E$ that coincides with $\widetilde{f}$ in the same set X.*

In spite of their appearances, Lemmas 4.2 and 4.3 are equivalent since there exists a homeomorphism φ from the rectangle $J \times I$ onto itself that takes the set X in the set $J \times 0$.

To prove this, we start by observing that it is enough to define a homeomorphism $\dot{\varphi}$ from the boundary of the rectangle to itself and extend it radially to the interior of the rectangle. In fact, if a is the center of the rectangle, every point $x \in J \times I$ can be written, in a unique way, as $x = (1-t)a + ty$, with $y \in \partial(J \times I) =$ the unique point of the boundary of $J \times I$ on the ray $\overrightarrow{ax}$. The radial extension of $\dot{\varphi}$ is defined by $\varphi(x) = (1-t)a + t\dot{\varphi}(y)$, where we suppose that $\dot{\varphi}(y)$ is known since $y \in \partial(J \times I)$.

Now, we define the homeomorphism $\dot{\varphi} \colon \partial(J \times I) \to \partial(J \times I)$ that takes $X = (J \times 0) \cup (\partial J \times I)$ onto $J \times 0$.

Figure 4.4 illustrates how the above homeomorphism $\dot{\varphi}$ operates: the points A, B, C, D, E, F are transformed respectively in A', B', C', D', E', F'. Each line segment in the boundary of the rectangle connecting two of these points is transformed by $\dot{\varphi}$, linearly, into the line segment that connects the corresponding image points.

The case $X = (J \times 0) \cup (s_1 \times I)$ is analogous.

Proof of Lemmas 4.2 and 4.3. Case 1. Suppose initially that $E = B \times F$ and $\pi \colon B \times F \to B$ is the projection onto the first factor. We have $\widetilde{a}(s) =$

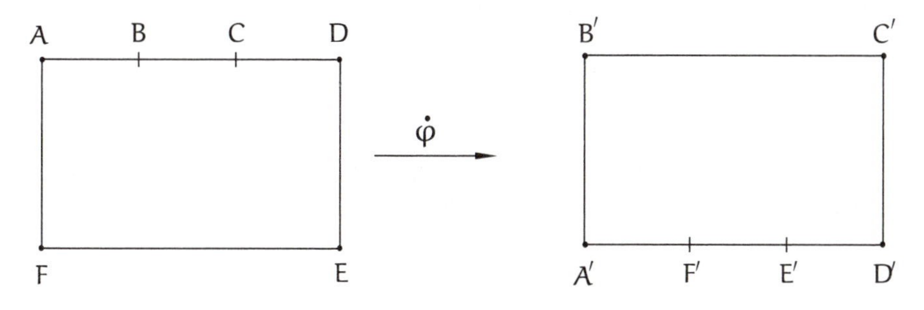

Figure 4.4.

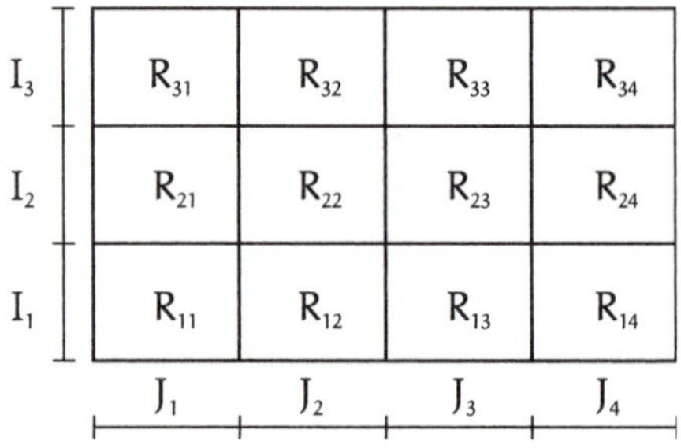

Figure 4.5.

$(a(s), b(s))$. We define $\widetilde{H}: J \times I \to B \times F$ by $\widetilde{H}(s,t) = (H(s,t), b(s))$. This proves Lemma 4.2 (and therefore 4.3) in this case.

Case 2. Now, suppose that there exists a global trivialization; that is, a homeomorphism $\varphi: B \times F \to E$, with $\pi(\varphi(x,y)) = x$. From the previous case, there exists a lifting $\widetilde{K}: J \times I \to B \times F$ of the homotopy H, starting with $\varphi^{-1} \circ \widetilde{a}$. Then $\widetilde{H} = \varphi \circ \widetilde{K}: J \times I \to E$ is a lifting of H, starting with $\widetilde{a}$, which proves Lemma 4.2 (and hence, Lemma 4.3) in the present case.

Case 3. (General) In the general case, we use the compactness of $J \times I$ to obtain a decomposition $J = J_1 \cup \ldots \cup J_k$, $I = I_1 \cup \ldots \cup I_r$ in consecutive intervals in such a way that each rectangle $R_{ij} = J_i \times I_j$ has its image $H(R_{ij})$ contained in a distinguished neighborhood $V_{ij} \subset B$, base of a local trivialization $\varphi_{ij}: V_{ij} \times F \to \pi^{-1}(V_{ij})$.

The map $\widetilde{H}$ is already specified on the left vertical face and the horizontal bottom face of R_{11}. From Case 2, we can extend it to the rectangle R_{11} in such a way to satisfy the condition $\pi \circ \widetilde{H} = H$. We use a similar argument with R_{12}, R_{13}, and so on until R_{1k} and, successively, in the following horizontal rows. (In the last rectangle of each horizontal row, $\widetilde{H}$ is already specified in the base and in the two vertical faces.) □

Proof of Proposition 4.6 Let $x_0 \in E$, $y_0 = \pi(x_0)$, and $F = \pi^{-1}(y_0)$. The inclusion map $i: F \to E$ induces a homomorphism

$$i_{\#}: \pi_1(F, x_0) \to \pi_1(E, x_0),$$

and we must show that it is surjective. For this, consider an arbitrary closed path $\widetilde{a}: I \to E$, with base point x_0. Then $a = \pi \circ \widetilde{a}$ is a closed path in B, with base y_0. Since B is simply connected, there exists a homotopy $H: a \cong e_{y_0}$. Define a continuous map $\widetilde{f}: X \to E$, with $X = (I \times 0) \cup (\partial I \times I)$ by

$$f(s,0) = \widetilde{a}(s), \widetilde{f}(0,t) = \widetilde{f}(1,t) = x_0, \ t \in I.$$

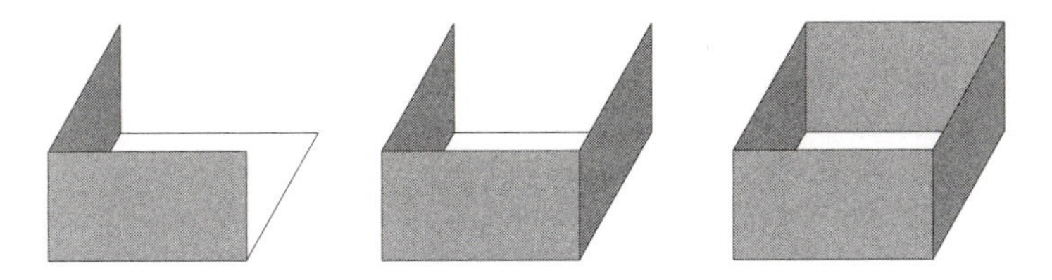

Figure 4.6. Three possible forms for X in Lemma 4.5.

The homotopy H coincides with $\pi \circ \widetilde{f}$ in X. By Lemma 4.3 there exists a homotopy $\widetilde{H}\colon I \times I \to E$ such that $\pi \circ \widetilde{H} = H$ and $\widetilde{H}$ coincides with $\widetilde{f}$ in X. This means that $\widetilde{H}$ is a homotopy between $\widetilde{a}$ and a closed path $\widetilde{b}\colon I \to E$ such that $\widetilde{b}(s) = \widetilde{H}(s,1)$ satisfies the condition $(\pi \circ \widetilde{b})(s) = (\pi \circ \widetilde{H})(s,1) = H(s,1) = y_0$, for every $s \in I$, hence b is a path in the fiber $F = \pi^{-1}(y_0)$. Therefore, every closed path $\widetilde{a}$ in E is homotopic to a path in F, which proves the surjectivity of the homomorphism $i_{\#}$. □

Now we prove Proposition 4.7. The proof is based on Lemmas 4.4 and 4.5 below, which are analogous to Lemmas 4.2 and 4.3 used in the proof of Proposition 4.6.

In the statements that follow, $R = J \times L$ is a rectangle, $R \times I$ a rectangular block and X is a subset of the boundary $\partial(R \times I)$, which can be the base $R \times 0$, or the union of this base with one, or more than one, vertical faces (see Figure 4.6).

Lemma 4.4. **(Homotopy lifting in rectangles)** *Let $\pi\colon E \to B$ be a locally trivial fibration. Given $\widetilde{g}\colon R \to E$ continuous, every homotopy $H\colon R \times I \to F$ that starts with $g = \pi \circ \widetilde{g}$ has a lifting $\widetilde{H}\colon R \times I \to E$ that starts with $\widetilde{g}$.*

Lemma 4.5. *Let $\pi\colon E \to B$ be a locally trivial fibration. Given $\widetilde{f}\colon X \to E$ continuous, every homotopy $H\colon R \times I \to B$ that coincides with $f = \pi \circ \widetilde{f}$ in X has a lifting $\widetilde{H}\colon R \times I \to E$ which coincides with $\widetilde{f}$ in the same set X.*

As in the previous case, Lemmas 4.4 and 4.5 are equivalent because there exists a homeomorphism φ from the rectangular block $R \times I$ onto itself that transforms the set X onto the set $R \times O$. In order to obtain φ, we define a homeomorphism $\dot{\varphi}$ from the boundary $\partial(R \times I)$ onto itself and extend it radially, by defining $\varphi((1-t)a + ty) = (1-t)a + t\dot{\varphi}(y)$, where a is the center of the block $R \times I$, $x = (1-t)a + ty$ is an arbitrary point of this block, and y is the point where the ray $\overrightarrow{ax}$ intersects the boundary $\partial(R \times I)$.

For $X = (R \times O) \cup (\partial R \times I)$ (that is, the base and the four vertical faces), the homeomorphism $\dot{\varphi}$ is described in Figure 4.7. Each point

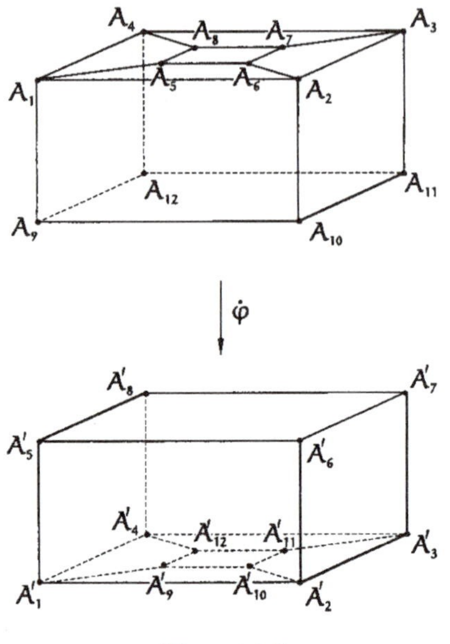

Figure 4.7.

$A_i (1 \leq i \leq 12)$ is transformed by $\dot{\varphi}$ in the point A'_i. The quadrilaterals, indicated in the figure, which have the points A_i as vertices, are transformed, by obvious homeomorphisms, in the quadrilaterals whose vertices are the corresponding points A'_i. The other options for X are handled in a similar way.

The proof of Lemma 4.4 (and hence of 4.5) is completely analogous to that of Lemmas 4.2 and 4.3, and for this reason we omit it.

Proof of Proposition 4.7. Let $x_0 \in E$, $y_0 = \pi(x_0)$, $F = \pi^{-1}(y_0)$ and $i\colon F \to E$ the inclusion map. It is enough to prove that the induced homomorphism $i_\#\colon \pi_1(F, x_0) \to \pi_1(E, x_0)$ is injective. Let $\widetilde{a}\colon I \to F$ be a closed path, with base in the point x_0. Suppose that, considered as a path in E, $\widetilde{a}$ is homotopic to a constant. (This means that $i_\#([\widetilde{a}]) = 0$.) Let $\widetilde{K}\colon I \times I \to E$ be the homotopy between $\widetilde{a}$ and the constant e_{x_0}. We use the notation $R = I \times I$, $X = (R \times 0) \cup (\partial R \times I)$ and define the continuous map $\widetilde{f}\colon X \to E$, by $\widetilde{f}(s,t,0) = \widetilde{K}(s,t)$, $\widetilde{f}(0,t,u) = \widetilde{f}(1,t,u) = \widetilde{f}(s,1,u) = x_0$, $\widetilde{f}(s,0,u) = \widetilde{a}(s)$ for s, t, $u \in I$. X consists of the base and the vertical faces of the block $R \times I$. In the base, $\widetilde{f}$ coincides with $\widetilde{K}$. In the lateral and posterior faces, f is constant, equal to x_0. In each horizontal segment of the previous face, f reproduces $\widetilde{a}$.

The map $K = \pi \circ \widetilde{K}\colon R \to B$ satisfies $K(\partial R) = y_0$. Since $\pi_2(B) = 0$, there exists a homotopy $H\colon R \times I \to B$ between K and the constant map $K \to y_0$, such that H is constant, equal to y_0, on every lateral face of the block $R \times I$ (therefore, $H = y_0$ on every faces of $R \times I$, except the base). Thus, H coincides with $f = \pi \circ \widetilde{f}$ in X. It follows from Lemma 4.5 that H has a lifting $\widetilde{H}\colon R \times I \to E$ that coincides with $\widetilde{f}$ in X. Since $H(s,t,1) = y_0$ for any s, $t \in I$, it follows that $\widetilde{H}(s,t,1) \in F$. Therefore, we can define a homotopy $H_1\colon I \times I \to F$ by $H_1(s,t) = \widetilde{H}(s,t,1)$. It is not difficult to prove that $H_1 : \widetilde{a} \cong e_{x_0}$ in F, hence $[\widetilde{a}] = 0$ in $\pi_1(F, x_0)$ and $i_\#$ is injective. □

In order to finish these considerations about the fundamental group of some groups of matrices, we study now the *simplectic group* $\mathrm{Sp}(n)$. We show that it is simply connected.

Let $\mathbb{H}$ be the field of quaternions. In the vector space $\mathbb{H}^n$, whose elements are ordered lists $v = (v_1, \ldots, v_n)$ of n quaternions, we define the inner product

$$\langle v, w \rangle = \sum_{r=1}^{n} v_r \cdot \overline{w}_r,$$

where the conjugate of the quaternion $w = t + xi + yj + zk$ is $\overline{w} = t - xi - yj - zk$. The quaternions v and w are called *orthogonal* when $\langle v, w \rangle = 0$.

The elements of the group $\mathrm{Sp}(n)$ are the $n \times n$ matrices whose columns (and rows) are pairwise orthogonal unit vectors in $\mathbb{H}^n$. For each $n \in \mathbb{N}$, $\mathrm{Sp}(n)$ is a bounded and closed set in $\mathbb{H}^n = \mathbb{R}^{4n}$, hence it is compact.

As in the cases of $\mathrm{SO}(n)$ and $\mathrm{SU}(n)$, the map

$$\pi\colon \mathrm{Sp}(n) \to S^{4n-1},$$

which associates to each simplectic matrix T its first column $\pi(T)$, is a locally trivial fibration, with typical fiber $\mathrm{Sp}(n-1)$, if $n \geq 2$.

When $n = 1$, we have the group $\mathrm{Sp}(1) = S^3$. It follows from Proposition 4.6 that $\mathrm{Sp}(n)$ is simply connected for every $n \geq 1$. For a description of $\mathrm{Sp}(n)$ without using quaternions, see Exercise 15.

4.3 Exercises

1. A quaternion commutes with $w = a + bi + cj + dk$ if, and only if, it commutes with $w' = bi + cj + dk$.

2. Determine all of the quaternions that commute with $u = 4 + 3i + 2j + k$ and, from this, describe the rotation axis of φ_u in $\mathbb{R}^3$.

3. With the exception of the negative real numbers, every non-null quaternion has two square roots. The square roots of the real quaternion $-a$, where $a \in \mathbb{R}^+$, fill up the sphere $S = \{xi + yj + zk; x^2 + y^2 + z^2 = a\}$.

4. Let $u' = ai + bj + ck$ and $v' = xi + yj + zk$ be the vector parts of the quaternions $u = d + ai + bj + ck$ and $v = t + xi + yj + zk$. Show that $u \cdot v - v \cdot u = 2u' \times v'$, where $u' \times v'$ represents the classical vector product.

5. Determine the matrix of the linear transformation $\varphi_u \colon \mathbb{R}^3 \to \mathbb{R}^3$ for the quaternion $u = a + bi + cj + dk \in S^3$.

6. Let $\varphi \colon S^3 \to \mathrm{SO}(3)$ be the homomorphism defined in Section 7. Show that, if $u = \cos\alpha + \sin\alpha \cdot i$ and $v = \cos\alpha + \sin\alpha \cdot k$, then φ_u and φ_v are the rotations with angles 2α around the axis i and k, respectively, in $\mathbb{R}^3$. Also, show that every rotation in $\mathbb{R}^3$ is the composition of at most three rotations, each one of them around one of these axes. Conclude from this a new proof that φ is surjective.

7. The Hopf fibration $h \colon S^3 \to S^2$ does not have a section $\tau \colon S^2 \to S^3$. (Suggestion: Suppose, by contradiction, that τ existed. Then $(x, u) \to u \cdot \tau(x)$ would be a homeomorphism between $S^2 \times S^1$ and S^3.)

8. Prove that a continuous map $f \colon S^2 \to S^2$ either has a fixed point or it has a point which is transformed into its antipode.

9. A homogeneous system of two linear equations with three variables:

$$a_1 x + b_1 y + c_1 z = 0$$
$$a_2 x + b_2 y + c_2 z = 0,$$

where the row vectors $\ell_1 = (a_1, b_1, c_1)$ and $\ell_2 = (a_2, b_2, c_2)$ are both non-null, always admits non trivial solutions $w = (x, y, z) \neq 0$. Is it possible to choose, for *every* pair of non-null vectors ℓ_1, $\ell_2 \in \mathbb{R}^3$, a non trivial solution w that depends continuously on the rows ℓ_1 and ℓ_2?

10. The matrices in the group $\mathrm{SU}(2)$ have the form

$$\begin{bmatrix} a & -\bar{b} \\ b & \bar{a} \end{bmatrix},$$

where a, $b \in \mathbb{C}$ and $|a|^2 + |b|^2 = 1$. Conclude that $\mathrm{SU}(2)$ is homeomorphic to S^3.

11. Let G be a Lie group, $H \subset G$ a closed subgroup, and $\pi \colon G \to G/H$ the standard projection. Prove that π is a locally trivial fibration. Conclude that if G and H are connected, then the fundamental group of G/H is abelian.

12. Let X be a simply connected space. Show that $\pi_2(X) = 0$ if, and only if, the space $\Omega(X, x_0)$ is simply connected. (See Exercise 3, Chapter 2.) In general, we define the *second homotopy group* of the space X by $\pi_2(X, x_0) = \pi_1(\Omega(X, x_0), x_0)$.

13. Show that the projective space P^2 admits no non-null continuous tangent vector field.

14. Let $B = B[0, \pi] \subset \mathbb{R}^3$ be the closed ball with center O and radius π in $\mathbb{R}^3$. Define a continuous surjection $\varphi\colon B \to \mathrm{SO}(3)$ by associating to each $x \in B$ the rotation $\varphi(x)$ of $|x|$ radians around the axis determined by x. Show that $\varphi(x) = \varphi(y)$ if, and only if, $|x| = |y| = 1$ and $y = -x$. By passing to the quotient, obtain again a homeomorphism $\overline{\varphi}\colon P^3 \to \mathrm{SO}(3)$.

15. Every quaternionic matrix W can be represented by $W = X + Y.j$, where X and Y are complex matrices.

a) Prove that the correspondence $W \mapsto f(W)$, where

$$f(W) = \begin{bmatrix} X & Y \\ -\overline{Y} & X \end{bmatrix},$$

is an isomorphism of the algebra of quaternionic $n \times n$ matrices onto a subalgebra of the complex $2n \times 2n$ matrices.

b) Prove that $W \in Sp(n)$; that is, $W \cdot W^* = I$ (where W^* is the conjugate of the transpose of W), if, and only if, $f(W) \cdot f(W)^* = I$; that is, $f(W)$ is unitary.

c) Prove that a complex $2n \times 2n$ matrix Z is of the form $f(W)$ if, and only if, $Z^T J Z = J$, where

$$J = \begin{bmatrix} 0 & -I \\ I & 0 \end{bmatrix}.$$

Conclude that $\mathrm{Sp}(n)$ can be identified with the group Z of $2n \times 2n$ complex matrices such that $Z^T J Z = J$.

16. Determine the fundamental group of the set of all $n \times n$ real matrices with determinant 1.

Chapter 5

The Winding Number

In this chapter, we use our results from the computation of the fundamental group of the circle S^1 in order to study the homotopy of closed plane curves in more detail.

5.1 The Winding Number of a Closed Plane Curve

We consider the paths $c\colon J \to X$, defined on a compact interval $J = [s_0, s_1]$, not necessarily the unit interval $I = [0, 1]$.

Let p be a point of the plane $\mathbb{R}^2$. Since $\mathbb{R}^2 - \{p\}$ has the same homotopy type of the circle S^1, its fundamental group is $\mathbb{Z}$. In particular, since $\mathbb{Z}$ is abelian, each element of the fundamental group $\pi_1(\mathbb{R}^2 - \{p\})$ may be considered as a free homotopy class of closed paths in $\mathbb{R}^2 - \{p\}$. Any of these classes, $\gamma = [c]$, is determined by an integer which measures the net number of turns of the path $c\colon I \to \mathbb{R}^2 - \{p\}$ around the point p.

Let $c\colon J \to \mathbb{R}^2 - \{p\}$ be a path in the plane whose image does not contain the point p. As we know, there exists a continuous function $\widetilde{c}\colon J \to \mathbb{R}$ such that

$$c(s) = p + \rho(s)e^{i\widetilde{c}(s)} \quad \text{for every} \quad s \in J,$$

where $\rho(s) = |c(s) - p|$. The function $\widetilde{c}$ is an angle function of the path

$$s \mapsto \frac{c(s) - p}{|c(s) - p|}$$

in S^1. It is determined up to an additive constant, which is an integral multiple of 2π.

Suppose that the path c is *closed*; that is, $c(s_0) = c(s_1)$. Then, for every angle function $\widetilde{c}$, the difference $\widetilde{c}(s_1) - \widetilde{c}(s_0)$ is an integral multiple of 2π, which does not depend on the choice of $\widetilde{c}$.

The *winding number* of the closed path $c\colon J \to \mathbb{R}^2 - \{p\}$ around the point p is the integer

$$n(c,p) = \frac{\widetilde{c}(s_1) - \widetilde{c}(s_0)}{2\pi}.$$

In other words, $n(c,p)$ is the degree of the path $a\colon J \to S^1$, defined by the radial projection

$$a(s) = \frac{c(s) - p}{|c(s) - p|}$$

of the path a onto the circle S^1.

The most important properties of the integer $n(c,p)$ are summarized in the proposition below. (As usually happens with propositions with long statements, its proof is very easy.)

Proposition 5.1. *The integer $n(c,p)$ has the following properties:*

1. *Let $c\colon [s_0, s_2] \to \mathbb{R}^2 - \{p\}$ be a path such that $c(s_0) = c(s_1) = c(s_2)$, where $s_0 < s_1 < s_2$. Define $c_1 = c|[s_0, s_1]$ and $c_2 = c|[s_1, s_2]$. Then $n(c,p) = n(c_1,p) + n(c_2,p)$.*

2. *The closed paths $c, c'\colon [s_0, s_1] \to \mathbb{R}^2 - \{p\}$ are free homotopic if, and only if, $n(c,p) = n(c',p)$.*

3. *If the points p and q can be connected by a path in the complement of the image $c(J)$ of the closed path $c\colon J \to \mathbb{R}^2 - \{p\}$, then $n(c,p) = n(c,q)$. In other words, $n(c,p)$ is, as a function of p, (keeping c fixed) constant in each connected component of $\mathbb{R}^2 - c(J)$.*

4. *Given the closed path $c\colon [s_0, s_1] \to \mathbb{R}^2 - \{p\}$, let $\varphi, \psi\colon [t_0, t_1] \to [s_0, s_1]$ be continuous functions such that $\varphi(t_0) = \psi(t_1) = s_0$ and $\varphi(t_1) = \psi(t_0) = s_1$. Then $n(c \circ \varphi, p) = n(c,p) = -n(c \circ \psi, p)$.*

5. *For every $k \in \mathbb{Z}$ and every $p \in \mathbb{R}^2$, the closed path $c\colon [0, 2\pi] \to \mathbb{R}^2 - \{p\}$, defined by $c(s) = p + e^{iks}$, satisfies $n(c,p) = k$.*

6. *Let $c, c'\colon J \to \mathbb{R}^2 - \{p\}$ be two closed paths such that, for every $s \in J$, the line segment $[c(s), c'(s)]$ does not contain the point p. Then $n(c,p) = n(c',p)$.*

7. *Let $c, c'\colon J \to \mathbb{R}^2 - \{0\}$ be two closed paths. If $|c(s) - c'(s)| < |c(s)|$ for every $s \in J$, then $n(c,0) = n(c',0)$.* (Theorem of Rouché.)

Proof. 1. Let $\widetilde{c}\colon [s_0, s_2] \to \mathbb{R}$ be an angle function for c. Then the restrictions $\widetilde{c}_1 = \widetilde{c}|[s_0, s_1]$ and $\widetilde{c}_2 = \widetilde{c}|[s_1, s_2]$ are angle functions for c_1 and c_2 respectively, and the result follows.

2. Since the map $h_p\colon \mathbb{R}^2 - \{p\} \to S^1$, defined by

$$h_p(z) = \frac{z - p}{|z - p|},$$

is a homotopy equivalence and $(h_p \circ c)(s) = e^{i\widetilde{c}(s)}$, we have $n(c, p) = n(h_p \circ c)$. Hence, $c \cong c' \Leftrightarrow h_p \circ c \cong h_p \circ c' \Leftrightarrow n(h_p \circ c) = n(h_p \circ c') \Leftrightarrow n(c, p) = n(c', p)$.

3. Let $a\colon I \to \mathbb{R}^2 - c(J)$ be a path such that $a(0) = p$ and $a(1) = q$. Define $H\colon J \times I \to S^1$ by

$$H(s, t) = \frac{c(s) - a(t)}{|c(s) - a(t)|}.$$

Then $H(s, 0) = h_p \circ c$ and $H(s, 1) = h_q \circ c$ for every $s \in J$. (We are using the notation from the previous item.) Hence H is a free homotopy between the closed paths $h_p \circ c, h_q \circ c\colon J \to S^1$. It follows that $n(c, p) = n(h_p \circ c) = n(h_q \circ c) = n(c, q)$.

4. This results from Proposition 2.2.

5. Obvious.

6. This follows from item 2 above, and Example 1.2.

7. This follows from the previous item. □

Example 5.1. Let $p\colon \mathbb{C} \to \mathbb{C}$ be the complex polynomial

$$p(z) = a_0 + a_1 z + \cdots + a_k z^k,$$

of degree $k > 0$. For every real number $r \geq 0$, p transforms the circle of center 0 and radius r (which degenerates in a point when $r = 0$) into a closed curve of the plane. Suppose that p does not have any root z with $|z| = r$. How many times does this closed curve turn around the origin? More precisely, let $c_r\colon [0, 2\pi] \to \mathbb{C} - \{0\}$ be the closed path defined by $c_r(s) = p(r \cdot e^{is})$. The problem consists in determining $n(c_r, 0)$. In general, this number depends on r.

We prove now that, for all sufficiently large r, $n(c_r, 0) = k$. (By the way, this is the origin of the name *degree* of a closed path in S^1.) This is easy when the polynomial p reduces to a monomial $p(z) = a_k z^k$. In the general case, we write $p(z) = a_k z^k + q(z)$, where $q(z)$ is polynomial of degree $\leq k - 1$. Hence,

$$|p(z) - a_k z^k| = |a_k z^k| \cdot f(z),$$

with

$$f(z) = \frac{q(z)}{a_k z^k};$$

therefore;

$$\lim_{|z|\to\infty} f(z) = 0.$$

Hence, there exists a positive real number r_0 such that

$$|z| = r > r_0 \Rightarrow |p(z) - a_k z^k| < |a_k z^k|.$$

Then, from Rouché's theorem, for every $r > r_0$, the closed path c_r turns k times around the origin. ◁

Later on, in Section 5.4 of this chapter, we will prove a sharper result than the one in the above example. Now we use the result from the example to give a proof of the famous Fundamental Theorem of Algebra. This is a very interesting application of the concept of winding number.

Theorem 5.1. (Fundamental Theorem of Algebra) *Every complex polynomial of degree $k > 0$ has at least one complex root.*

Proof. Suppose, by contradiction, that $p(z) \neq 0$ for every z. Then $c_r\colon [0, 2\pi] \to \mathbb{C} - \{0\}$ is defined for every $r \geq 0$. Note that for any two non-negative real numbers r, r' we have $c_r \simeq c_{r'}$, using the homotopy $H(s,t) = p(((1-t)r + tr')e^{is})$. Therefore $n(c_r, 0)$ does not depend on r. Now, $n(c_0, 0) = 0$ because c_0 is constant. As we have just proved in Example 5.1, $n(c_r, 0) = k$ for r sufficiently large. This is a contradiction; therefore, we must have $p(z) = 0$ for some z. □

The proof given above is the third of the four proofs devised by Gauss for the Fundamental Theorem of Algebra. (The first was his doctor's thesis.)

5.2 The Graustein-Whitney Theorem

In this section, we introduce the concept of regular homotopy and use the winding number in order to provide a necessary and sufficient condition for the regular homotopy of two closed regular paths. This problem is closely related with the well-known problem of sphere eversion.

We say that a path $a\colon [s_0, s_1] \to \mathbb{R}^2$ is *regular* when it is of class C^1 and $a'(s) \neq 0$ for all $s \in [s_0, s_1]$. A regular path $a\colon [s_0, s_1] \to \mathbb{R}^2$ such that $a(s_0) = a(s_1)$ and $a'(s_0) = a'(s_1)$ will be called a *regular curve.* Since it is a closed path, a regular curve $a\colon [s_0, s_1] \to \mathbb{R}^2$ defines an immersion of S^1

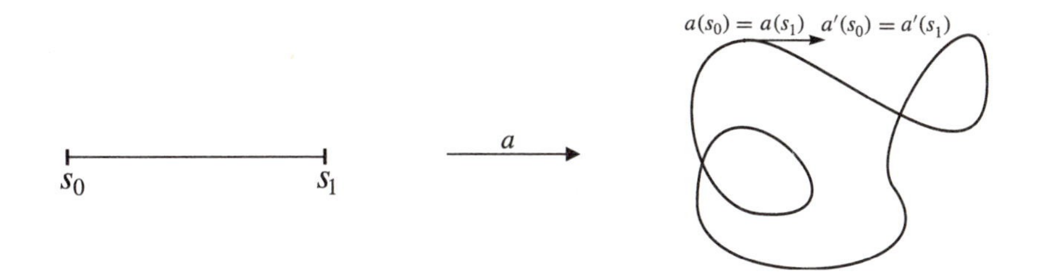

Figure 5.1. A regular curve.

into $\mathbb{R}^2$; therefore, although it may have self intersections, for each value of the parameter s, it has a well defined (unique) tangent line at $a(s)$ whose direction is given by the nonzero vector $a'(s)$ (see Figure 5.1).

The *rotation number* of the regular curve $a\colon [s_0, s_1] \to \mathbb{R}^2$ is, by definition, the winding number $n(a') = n(a', 0)$ of the derivative path $a'\colon [s_0, s_1] \to \mathbb{R}^2 - \{0\}$ around the origin $0 \in \mathbb{R}^2$.

More explicitly, let $\theta\colon [s_0, s_1] \to \mathbb{R}$ be an angle function for a'; that is,

$$a'(s) = |a'(s)|e^{i\theta(s)}$$

for all $s \in [s_0, s_1]$. Then the rotation number of the regular curve a is the integer

$$n(a') = \frac{\theta(s_1) - \theta(s_0)}{2\pi}.$$

If $\varphi\colon [t_0, t_1] \to [s_0, s_1]$ is a C^1 homeomorphism with positive derivative, and $\varphi'(t_0) = \varphi'(t_1)$, then $b = a \circ \varphi\colon [t_0, t_1] \to \mathbb{R}^2$ is again a regular curve, which we call a *reparametrization* of a. The reparametrized curve $b = a \circ \varphi$ has the same rotation number as a. To see this, observe that for all $t \in [t_0, t_1]$,

$$b'(t) = (a \circ \varphi)'(t) = \varphi'(t)a'(\varphi(t)) = |\varphi'(t)a'(\varphi(t))|e^{i\theta(\varphi(t))} = |b'(t)|e^{i\theta(\varphi(t))},$$

so $\theta \circ \varphi\colon [t_0, t_1] \to \mathbb{R}$ is an angle function for b'. Thus,

$$n(b') = \frac{\theta(\varphi(t_1)) - \theta(\varphi(t_0))}{2\pi} = \frac{\theta(s_1) - \theta(s_0)}{2\pi} = n(a').$$

In particular, if we parametrize the regular curve a by arc length, its rotation number will not change. In this case, we have $a\colon [0, L] \to \mathbb{R}^2$, with $|a'(s)| = 1$ for all $s \in [0, L]$, where L is the length of a.

Let $a, b\colon [s_0, s_1] \to \mathbb{R}^2$ be regular curves. We wish to define the concept of regular homotopy $H(s, t)$ between a and b. It is natural to demand H to be a free homotopy between the two closed paths a and b and, moreover,

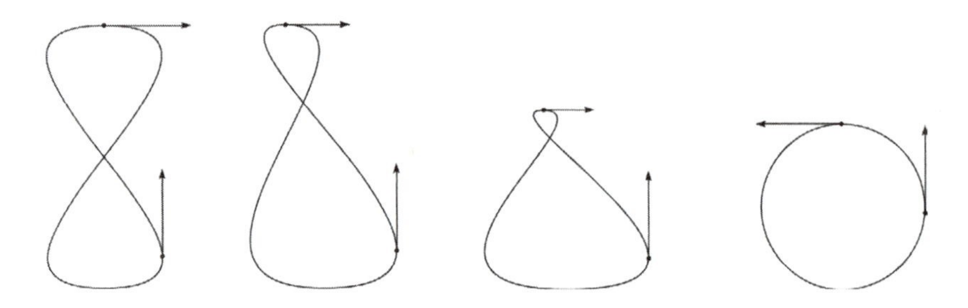

Figure 5.2. Non regular homotopy.

that each intermediate path $H_t(s)$ of the homotopy also be a regular curve. Consider the homotopy between the figure eight and the circle shown in Figure 5.2. Note that the derivative $\partial H/\partial s(s,t)$ is discontinuous at the point $(s_0, 1)$ where $a(s_0)$ is the highest point on the figure. Thus, this should not be considered a regular homotopy between the two regular curves.

Now we state precisely the concept of regular homotopy. Let $a, b\colon [s_0, s_1] \to \mathbb{R}^2$ be regular curves. A *regular homotopy* between a and b is a C^1 map $H\colon [s_0, s_1] \times I \to \mathbb{R}^2$ such that, for all $s \in [s_0, s_1]$ and $t \in I$:

1. $H(s, 0) = a(s), H(s, 1) = b(s)$;
2. $\frac{\partial H}{\partial s}(s, t) \neq 0$;
3. $H(s_0, t) = H(s_1, t)$ and $\frac{\partial H}{\partial s}(s_0, t) = \frac{\partial H}{\partial s}(s_1, t)$.

Condition 1 is the usual free homotopy property; Condition 2 guarantees the regularity of each intermediary path H_t, and Condition 3 guarantees that each intermediary path H_t is in fact a regular curve. Figure 5.3 shows two examples of regular homotopies.

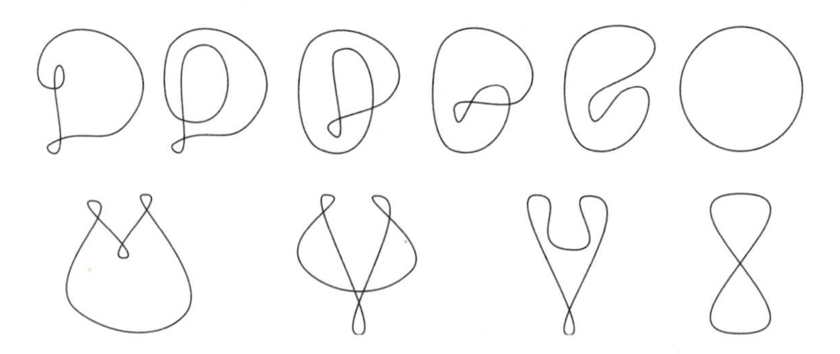

Figure 5.3. Regular homotopies.

Now we state the main result of this section:

Theorem 5.2. (Graustein-Whitney) *Two regular curves in the plane are regularly homotopic if, and only if, they have the same rotation number.*

In the proof, we need the following lemma.

Lemma 5.1. *Let $f\colon [0,L] \to S^1$ be a path. If its mean value*

$$m = \frac{1}{L}\int_0^L f(s)ds$$

belongs to S^1, then f is constant.

Proof. Let $g\colon [0,L] \to \mathbb{R}^2$ be a primitive of f; that is, $g'(s) = f(s)$ for all $s \in [0,L]$. Since $|g'(s)| = |f(s)| = 1$ for all s, g is a C^1 path parametrized by arc length. Now $|m| = 1$ implies $|g(L) - g(0)| = L$, so the length

$$L = \int_0^L |g'(s)|ds$$

equals the distance from the initial point $g(0)$ to the endpoint $g(L)$. The definition of arc length as the supremum of the lengths of the polygons inscribed in the path implies then that the image of g is the straight line segment from $g(0)$ to $g(L)$. If u is the unit vector on this segment, we have then $g(s) = g(0) + \varphi(s).u$. Since $|g'(s)| = |\varphi'(s)| = 1$, we must have $\varphi'(s) = \pm 1$. In fact, $\varphi'(s) = 1$ so $f(s) = g'(s) = u$. In other words, f is constant. □

Proof of the Graustein-Whitney Theorem. The "only if" part is easy: If H is a regular homotopy between the regular curves a, b then $\partial H/\partial s$ is a free homotopy between the paths $a', b'\colon [s_0, s_1] \to \mathbb{R}^2 - \{0\}$. Hence, a' and b' have the same winding number; that is, a and b have the same rotation number.

Now we prove the "if" part. In order to set the stage, we observe that, by means of a regular homotopy, we may alter the length of a regular curve to an arbitrary value. Hence, there is no loss of generality in assuming that we have two regular curves $a, b\colon [0,L] \to \mathbb{R}^2$, with the same length L, with $n(a') = n(b') = n$ and $|a'(s)| = |b'(s)| = 1$ for all $s \in [0,L]$, that is, both curves are parametrized by arc length.

Moreover, we may assume that $a(0) = b(0) = 0$ and $a'(0) = b'(0) = e_1$ (the unit vector of the x-axis). Therefore,

$$a'(s) = e^{i\alpha(s)} \quad \text{and} \quad b'(s) = e^{i\beta(s)},$$

where $\alpha, \beta\colon [0, L] \to \mathbb{R}$ are angle functions, with $\alpha(0) = \beta(0) = 0$ and $\alpha(L) = \beta(L) = 2\pi n$.

We define a linear homotopy $K\colon [0, L] \times I \to \mathbb{R}$ between α and β, given by $K(s,t) = (1-t)\alpha(s) + t\beta(s)$. With the help of K, we define a homotopy $H\colon [0, L] \times I \to \mathbb{R}^2$ between the curves a and b by setting

$$H(s,t) = \int_0^s e^{iK(u,t)} du - \frac{s}{L} \int_0^L e^{iK(u,t)} du.$$

The negative summand above is introduced in order to guarantee that all paths H_t, $0 \leq t \leq 1$, are closed. Next, we verify the details.

Since

$$a(s) = \int_0^s e^{i\alpha(u)} du \quad \text{and} \quad b(s) = \int_0^s e^{i\beta(u)} du,$$

it is clear that $H(s,0) = a(s)$ and $H(s,1) = b(s)$ for all $s \in [s_0, s_1]$. Moreover, it is easy to verify that $H(0,t) = H(L,t) = 0$ and

$$\frac{\partial H}{\partial s}(0,t) = \frac{\partial H}{\partial s}(L,t)$$

for all $t \in I$.

It remains to show that $\frac{\partial H}{\partial s}(s,t) \neq 0$ for all $s \in [s_0, s_1]$ and all $t \in I$. We have

$$\frac{\partial H}{\partial s}(s,t) = e^{iK(s,t)} - \frac{1}{L} \int_0^L e^{iK(u,t)} du.$$

The question we face is this: Given a path $\lambda\colon [0,1] \to S^1$, $\lambda(s) = e^{iK(s,t)}$, can its mean value

$$\frac{1}{L} \int_0^L \lambda(u) du$$

be a point in the circle S^1? By Lemma 5.1 this can only happen when λ is constant.

In our case, we conclude that H is a regular homotopy between a and b, provided it does not happen that, for some $t \in I$, the path $\lambda(s) = e^{iK(s,t)}$ or (which means the same) the function $\mu(s) = K(s,t)$ be constant. Since $\mu(0) = 0$, such constant should be zero, so that $0 = \mu(L) = K(L,t) = 2\pi n$. Hence $n = n(a') = n(b') = 0$ and, for some $t \in [0,1]$, the equality $(1-t)\alpha(s) + t\beta(s) = 0$ holds; that is,

$$\alpha(s) = \frac{t}{t-1}\beta(s),$$

for all $s \in [o, L]$. (Note that $0 < t < 1$ because α and β cannot be identically zero.)

Summing up: If the regular curves a, b have the same rotation number then the map H defined above is a regular homotopy between them, except when the following holds:

1. The rotation numbers of a and b are equal to zero;
2. $a'(s) = e^{i\alpha(s)}$, $b'(s) = e^{i\beta(s)}$ for all $s \in [0, L]$, where $\alpha, \beta \colon [0, L] \to \mathbb{R}$ are continuous functions both equal to zero at the endpoints 0 and L, with $\alpha = t\beta/(t-1)$ for some $t \in (0, 1)$. (That is, α is a negative constant multiple of β.)

To dispose of this remaining case, we take any continuous function $\gamma \colon [0, L] \to \mathbb{R}$ which is zero at the endpoints 0 and L and is not a constant multiple of β (nor of $\alpha = t\beta/(t-1)$, of course). Then we redefine $K \colon [0, L] \times I \to \mathbb{R}$ as

$$K(s, t) = (1-t)\alpha(s) + t\beta(s) + t(t-a)\gamma(s).$$

Since the last summand above is zero for $t = 0$ and $t = 1$, K is still a homotopy between α and β. However, for no value of $t \in [0, 1]$ the function $\mu(s) = K(s, t)$ vanishes identically.

This concludes the proof of the Graustein-Whitney theorem. □

Example 5.2. The circle and the figure eight (see Figure 5.2) are not regularly homotopic. In fact, the circle has rotation number ± 1 (depending on the orientation) and the rotation number of the figure eight is zero. ◁

The rotation number (Umlaufzahl) of a regular curve is a classical notion. It is the object of a famous theorem, known as the Umlaufsatz, according to which the rotation number of a simple (i.e., non-self-intersecting) regular curve is ± 1. This theorem has a very elegant proof, given in 1935 by H. Hopf, Hopf (1935), which can be found in doCarmo (1976), page 396. Two years after Hopf's proof, H. Whitney published the paper, Whitney (1937), in which he proved that two regular curves in the plane are regularly homotopic if, and only if, they have the same rotation number. In that paper, Whitney states that this result, with a simple proof, had been communicated to him by his colleague W. Graustein. Besides proving that, Whitney gave a method for computing the rotation number, by counting the algebraic number of times that the curve intersects itself. Note that, by using Hopf's theorem and the Graustein-Whitney theorem, we conclude that every simple regular curve is regularly homotopic to the circle S^1.

5.2.1 About Eversions

Consider a regular curve $a \colon [s_0, s_1] \to \mathbb{R}^2$ in the plane, and assume that it is *simple* (i.e., it has no self-intersections) and parametrized by arc length.

Thus, a is an embedding of the circle S^1 into the plane such that the velocity vector $a'(s)$ has norm 1. A *normal vector field* to a is a continuous nonzero vector field $n\colon [s_0, s_1] \to \mathbb{R}^2$, such that $\langle n(s), a'(s)\rangle = 0$ for all $s \in [s_0, s_1]$. The simple regular curve a always admits two unit normal vector fields. The choice of one of these unit normal vector fields is called an *orientation* of the curve. The orientation for which the basis $\{a'(s), n(s)\}$ has the same orientation of the Euclidean plane is called positive, and the other is the negative orientation. The curve a with positive orientation is denoted by a^+, and a^- denotes the curve with negative orientation. An *eversion* of a is a regular homotopy from a^+ to a^-. Thus an eversion of a simple regular curve has the effect of turning the curve inside out.

Example 5.3. (Circle eversion) In the case of the circle $S^1 \subset \mathbb{R}^2$, an eversion is a regular homotopy between the standard inclusion map $i\colon S^1 \to \mathbb{R}^2$, $i(p) = p$, and the antipodal map $\alpha\colon S^1 \to \mathbb{R}^2$, $\alpha(p) = -p$. The theorem of Graustein-Whitney shows that this eversion is impossible because the circle with positive orientation has rotation number $+1$ and the circle with negative orientation has rotation number -1. Figure 5.4 shows a natural attempt to obtain such an eversion. Note the sudden change (discontinuity) of the two horizontal normals at the end of the homotopy. As in the case of Figure 5.2, this indicates discontinuities of $\partial H/\partial s$ at $t = 1$, so the homotopy is not regular. $\triangleleft$

After the results from Hopf and Whitney, the subject of regular homotopies was studied again, twenty years later, by S. Smale in his Ph.D. thesis. Note that there are two possible ways to obtain generalizations of the Graustein-Whitney theorem: by taking instead of the plane $\mathbb{R}^2$ an arbitrary Riemannian manifold, or by replacing the regular curves $a, b\colon S^1 \to \mathbb{R}^2$ by immersions $a, b\colon S^k \to \mathbb{R}^n$. Smale studied both possibilities.

In the first case, he showed that the regular homotopy classes on a Riemannian manifold are in a $1-1$ correspondence with homotopy classes of loops in the unit tangent bundle of the manifold, Smale (1958).

In the second case, Smale extended his methods in order to study regular

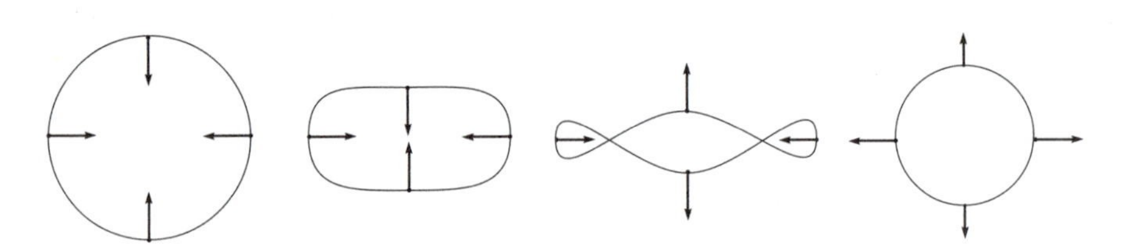

Figure 5.4. Attempt to evert the circle.

homotopies of immersed spheres into the Euclidean space Smale (1959). A consequence of his fundamental result is that *any two immersions of S^2 in $\mathbb{R}^3$ are regularly homotopic.* In the early days of his career (1957), Smale spoke of his work at a meeting at the University of Chicago. In the audience was the renowned topologist S. Eilenberg, who remarked:

> "This cannot be right because it implies that the antipodal map $\alpha\colon S^2 \to \mathbb{R}^3$ is regularly homotopic to the natural inclusion $i\colon S^2 \to \mathbb{R}^3$. In other words, according to you one can evert a sphere in 3-space, which is absurd".

Smale just smiled and replied:

> "I do not know how to figure geometrically the deformation. But I know that it can be done because my proof is correct".

Soon afterwards, A. Shapiro and B. Morin showed, by means of different ingenious devices, how to exhibit an explicit geometric regular deformation that turns the sphere S^2 inside out. The episode of Neptune's discovery was repeated. Morin's construction is particularly impressive because he is blind. For a detailed description of Shapiro's eversion, the reader should consult Francis & Morin (1979) or Phillips (1966). A more detailed history of the problem can be found in Chapter 6 of Francis (1987). Another good source is Levy (1995).

More recently, a new technique to construct an eversion of the sphere S^2 was described by W. Thurston. This technique provides more geometric insight than the others cited above. Its description can be found in Levy (1995). Thurston's eversion has been beautifully illustrated in the computer graphics video "Outside In", Levy *et al.* (1995), which was produced at the Geometry Center in 1994.

5.3 The Winding Number as a Curvilinear Integral

A *differential form* (of degree 1) in an open set $U \subset \mathbb{R}^2$ is an expression of the type

$$\omega = f dx + g dy,$$

where $f, g\colon U \to \mathbb{R}$ are functions of class C^1 in U; that is, they have continuous partial derivatives at every point of U.

The symbols dx and dy are interpreted as follows: Whenever we restrict the form ω to a path $c\colon J \to U$, of class C^1, with $c(t) = (x(t), y(t))$, we substitute dx by $x'(t)dt$ and dy by $y'(t)dt$.

Two forms, $\omega = f dx + g dy$ and $\omega_1 = f_1 dx + g_1 dy$, defined in the same open set $U \subset \mathbb{R}^2$, are *equal* when $f = f_1$ and $g = g_1$. The *sum* of the forms

ω and ω_1 is the form

$$\omega + \omega_1 = (f + f_1)dx + (g + g_1)dy.$$

The *product* of the form ω by the function $h\colon U \to \mathbb{R}$ is the form $h\omega = hf \cdot dx + hg \cdot dy$. We do not define multiplication of differential forms.

An example is provided by the *differential* $d\varphi$ *of a function* $\varphi\colon U \to \mathbb{R}$, of class C^2. In this case, we have the form

$$d\varphi = \frac{\partial \varphi}{\partial x}dx + \frac{\partial \varphi}{\partial y}dy,$$

still defined in the open set U which is the domain of φ.

Suppose that ω is a form in U. If there exists $\varphi\colon U \to \mathbb{R}$, of class C^2, such that $\omega = d\varphi$, we say that φ is an *exact* form.

Let $\omega = fdx + gdy$ be a differential form in the open set $U \subset \mathbb{R}^2$ and $c\colon J \to U$ a path of class C^1 in U, defined by $c(t) = (x(t), y(t)), t \in J$, where $x, y\colon J \to \mathbb{R}$ are continuous differentiable functions.

The (curvilinear) *integral of the form* ω *along the path* c is defined by

$$\int_c \omega = \int_{t_0}^{t_1} [f(x(t), y(t)) \cdot x'(t) + g(x(t), y(t)) \cdot y'(t)]dt, J = [t_0, t_1].$$

In the above expression, the integral on the right is the usual integral of a continuous function defined on the compact interval $[t_0, t_1]$ of real numbers.

Consider a differential form ω defined in an open set $U \subset \mathbb{R}^2$ and a path $c\colon J \to U$, of class C^1, where $J = [t_0, t_1]$. The two following properties are easy to prove.

1. Let $J = J_1 \cup J_2$ be the union of two compact intervals with a common endpoint. If $c_1 = c|J_1$ and $c_2 = c|J_2$ we have

$$\int_c \omega = \int_{c_1} \omega + \int_{c_2} \omega.$$

2. Let $\varphi, \psi\colon [s_0, s_1] \to [t_0, t_1]$ be functions of class C^1, with $\varphi(s_0) = \psi(s_1) = t_0$ and $\varphi(s_1) = \psi(s_0) = t_1$. Then

$$\int_{c \circ \varphi} \omega = \int_c \omega \quad \text{and} \quad \int_{c \circ \psi} \omega = -\int_c \omega.$$

We represent the path $c \circ \psi$ by $-c$.

It is possible to define $\int_c \omega$ even when the path $c\colon J \to U$ is only piecewise of class C^1; that is, we have $J = J_1 \cup J_2 \cup \ldots \cup J_k$, where each J_i is a compact interval that has exactly one point in common (one of the two endpoints) with J_{i+1}, and, taking $c_i = c|J_i (i = 1, \ldots, k)$, each c_i is of class C^1. In this case, we define:

$$\int_c \omega = \int_{c_1} \omega + \int_{c_2} \omega + \cdots + \int_{c_k} \omega.$$

As an example, we compute the integral $\int_c \omega$ in the case where $\omega = d\varphi$ is an exact form, differential of the function $\varphi\colon U \to \mathbb{R}$, of class C^2, and $c\colon J \to U$ is a path of class C^1. The derivative of the composite function $\varphi \circ c\colon J \to \mathbb{R}$, by the chain rule, is:

$$(\varphi \circ c)'(t) = \frac{d}{dt}\varphi(x(t), y(t)) = \frac{\partial \varphi}{\partial x} x' + \frac{\partial \varphi}{\partial y} y'.$$

It follows that

$$\int_c \omega = \int_c d\varphi = \int_{t_0}^{t_1} \left(\frac{\partial \varphi}{\partial x} x' + \frac{\partial \varphi}{\partial y} y' \right) dt =$$
$$= \int_{t_0}^{t_1} (\varphi \circ c)'(t) dt = \varphi(c(t_1)) - \varphi(c(t_0)).$$

In particular, it follows from the above computations that if $\omega = d\varphi$ is an exact form, the integral $\int_c \omega$ does not depend on the path c but only on the endpoints $c(t_0)$ and $c(t_1)$.

The statement that $\int_c \omega$ depends only on the endpoints of the path c in U is equivalent to stating that $\int_c \omega = 0$ for every closed path c in U.

A straightforward use of the definition shows that the above results hold, more generally, for paths c that are piecewise C^1.

A differential form $\omega = f dx + g dy$, in an open set $U \subset \mathbb{R}^2$, is called *closed* when

$$\frac{\partial f}{\partial y} = \frac{\partial g}{\partial x}$$

in U.

For example, if ω is exact, say $\omega = d\varphi$, then

$$f = \frac{\partial \varphi}{\partial x} \quad \text{and} \quad g = \frac{\partial \varphi}{\partial y},$$

hence,

$$\frac{\partial f}{\partial y} = \frac{\partial^2 \varphi}{\partial y \partial x} = \frac{\partial^2 \varphi}{\partial x \partial y} = \frac{\partial g}{\partial x}.$$

Therefore, every exact form is closed.

But not every closed form is exact. In fact, consider the differential form

$$w = \frac{-y}{x^2 + y^2} dx + \frac{x}{x^2 + y^2} dy,$$

defined on the open set $U = \mathbb{R}^2 - \{0\}$. We have

$$f = \frac{-y}{x^2 + y^2} \quad \text{and} \quad g = \frac{x}{x^2 + y^2}.$$

Hence,

$$\frac{\partial f}{\partial y} = \frac{\partial g}{\partial x} = \frac{y^2 - x^2}{(x^2 + y^2)^2}.$$

Therefore, ω is a closed form. Nevertheless, ω is not exact in U. More precisely, there does not exist $\varphi \colon U \to \mathbb{R}$, of class C^2, such that $\omega = d\varphi$. In fact, if this were true, the integral of ω along any closed piecewise C^1 path, contained in U, would be zero. But this does not happen, as we prove in the proposition below.

Proposition 5.2. *Let $c \colon J \to \mathbb{R}^2 - \{0\}$ be a closed piecewise C^1 path. Then*

$$n(c, 0) = \frac{1}{2\pi} \int_c \omega, \quad \text{where} \quad \omega = \frac{-y}{x^2 + y^2} dx + \frac{x}{x^2 + y^2} dy.$$

Proof. We suppose that c is of class C^1 and leave to the reader the general case. Let $\theta \colon J \to \mathbb{R}$ be an angle function for c. By taking $c(t) = (x(t), y(t))$, we have $x = \rho \cos\theta$, $y = \rho \sin\theta$, where $\rho^2 = x^2 + y^2$. (We are abbreviating $x = x(t)$, $y = y(t)$, $\rho = \rho(t)$, $\theta = \theta(t)$.) Taking derivatives, we obtain

$$x' = \rho' \cos\theta - \rho \sin\theta \cdot \theta'$$
$$y' = \rho' \sin\theta + \rho \cos\theta \cdot \theta'.$$

Making obvious substitutions, we have:

$$\int_c \omega = \int_{t_0}^{t_1} \left[\frac{-y}{x^2 + y^2} x' + \frac{x}{x^2 + y^2} y' \right] dt =$$
$$= \int_{t_0}^{t_1} \theta'(t) dt = \theta(t_1) - \theta(t_0) = 2\pi \cdot n(c, 0),$$

which concludes the proof. □

The above proposition expresses the winding number of a closed piecewise C^1 path around the origin of $\mathbb{R}^2$ as the integral of a certain differential form ω.

The admiration the reader might have for the person who guessed the form ω will decrease substantially if he notices that, if $\theta(x, y)$ is a determination of the angle between the vector $v = (x, y) \neq 0$ with the positive

x-axis, then

$$\cos\theta = \frac{x}{\sqrt{x^2+y^2}};$$

hence,

$$\theta(x,y) = \arccos\frac{x}{\sqrt{x^2+y^2}}.$$

An elementary computation shows that

$$d\left(\arccos\frac{x}{\sqrt{x^2+y^2}}\right) = \omega.$$

We remark emphatically that this does not mean that ω is exact in $U = \mathbb{R}^2 - \{0\}$, because the "function"

$$\theta(x,y) = \arccos\frac{x}{\sqrt{x^2+y^2}}$$

is not well defined in U. Two determinations $\theta_1(x,y)$ and $\theta_2(x,y)$ of the angle between $v = (x,y)$ and the semi-axis $0x$ differ locally by a constant (an integer multiple of 2π). This is why the differential $\omega = d\theta$ is well defined, although θ is not.

The differential form ω, defined in $\mathbb{R}^2 - \{0\}$, is called the *angle element* of the plane.

5.3.1 The Winding Number as a Complex Integral

The number of turns $n(c,0)$ of a closed piecewise C^1 path $c\colon J \to \mathbb{R}^2 - \{0\}$ around the origin can also be expressed by a complex integral:

$$n(c,0) = \frac{1}{2\pi i}\int_c \frac{dz}{z}.$$

In the above integral, we have

$$dz = dx + idy \quad \text{and} \quad \frac{1}{z} = \frac{1}{x+iy} = \frac{x-iy}{x^2+y^2}.$$

Hence,

$$\frac{dz}{z} = \frac{(x-iy)(dx+idy)}{x^2+y^2} = \omega_1 + i\omega_2$$

where

$$\omega_1 = \frac{x}{x^2+y^2}dx + \frac{y}{x^2+y^2}dy$$

and

$$\omega_2 = \frac{-y}{x^2+y^2}dx + \frac{x}{x^2+y^2}dy.$$

As it can be easily proved, $\omega_1 = d(\log\sqrt{x^2+y^2})$ is an exact form in $\mathbb{R}^2 - \{0\}$; therefore, $\int_c \omega_1 = 0$ for every closed piecewise C^1 path c, in $\mathbb{R}^2 - \{0\}$. On the other hand, ω_2 is the angle element of the plane. Hence,

$$\int_c \frac{dz}{z} = i\int_c \omega_2 = 2\pi i \cdot n(c,0).$$

5.4 Winding Number and Polynomial Roots

In this section, we use the winding number to obtain information about the roots of a complex polynomial. In fact, we prove a sharper result than the one in Example 5.1:

Given the polynomial p, for each $r > 0$ such that p does not have a root of modulus r, the number $n(c_r, 0)$ is equal to the number of roots of p within the disk of center O and radius r, each root being counted according to its multiplicity.

We start with the change of variable formula for line integrals in the plane.

Let $U, V \subset \mathbb{R}^2$ be open sets and $F\colon U \to V$ be a C^1 map, that is, $F(x,y) = (f(x,y), g(x,y))$, where f and g are continuously differentiable functions. If $\omega = adx + bdy$ is a differential form in V, with coefficients $a, b\colon V \to \mathbb{R}$, and $c\colon J \to U$ is a piecewise C^1 path, then $F \circ c\colon J \to V$ is a piecewise C^1 path in V and the change of variable formula is

$$\int_{F\circ c} adx + bdy = \int_c (a\circ F)df + (b\circ F)dg.$$

If we write $F^*\omega = (a\circ F)df + (b\circ F)dg$, the above formula becomes

$$\int_{F\circ c} \omega = \int_c F^*\omega.$$

For instance, when $c(s) \neq 0$ and $F(c(s)) \neq 0$ for all $s \in J$, the change of variable formula for the angle element reads

$$\int_{F\circ c} \frac{xdy - ydx}{x^2+y^2} = \int_c \frac{fdg - gdf}{f^2+g^2}.$$

The proof of the change of variable formula is straightforward, so we omit it.

Let $a \in U$ be such that $F(a) = 0$ but $F(z) \neq 0$ for all $z \neq a$ in a disk D of center a and radius r contained in U. Then we say that a is an *isolated zero* of the map F. The path $c\colon [0, 2\pi] \to U$, given by $c(s) = r + r.e^{is}$, is such that

$$n(F \circ c, 0) = \frac{1}{2\pi} \int_{F \circ c} \frac{xdy - ydx}{x^2 + y^2}$$

is the net number of turns of the path $F \circ c\colon s \mapsto F(c(s))$ around 0. The change of variable formula says that this number equals

$$\frac{1}{2\pi} \int_c \frac{fdg - gdf}{f^2 + g^2}.$$

It is called the *local degree* of F at the isolated zero a.

Let us look at the special case in which $F\colon \mathbb{C} \to \mathbb{C}$ is given by $F(z) = (z - a)^m$. We claim that the local degree of F at a equals m.

In fact, if $c(s) = a + r.e^{is}$ then $F(c(s)) = r^m.e^{ims}$ so, when s varies from 0 to 2π, the point $F(c(s))$ covers m times, in the positive sense, the circle of radius r^m and center 0.

In the same fashion, we see that, if b is any non zero complex number, the local degree of $F(z) = b.(z - a)^m$ at a is again m.

Now let $p(z)$ be an arbitrary polynomial. The complex number a is called a *root of multiplicity* m of $p(z)$ when

$$p(z) = (z - a)^m.q(z),$$

where $q(z)$ is a polynomial such that $q(a) \neq 0$.

When a is a root of multiplicity m of the polynomial $p(z)$, we claim that, considered as a map $p\colon \mathbb{C} \to \mathbb{C}$, the polynomial $p(z)$ has local degree m at a. In other words, when $z = r + e^{is}$, $0 \leq s \leq 2\pi$, describes once, in the positive sense, the boundary of a disk with center a and radius r that contains no other root of $p(z)$ but a, then $p(z)$ runs m times around 0 in the counterclockwise.

To see this, we let $q(a) = b$, so $q(z) = b + \varphi(z)$, with $\varphi(a) = 0$. Take $r > 0$ so small that $|b.\varphi(z)| < 1$ for all z such that $|z - a| \leq r$, Then

$$|z - a| = r \Rightarrow p(z) = (z - a)^m q(z) = b.(z - a)^m + b.\varphi(z).(z - a)^m,$$

with

$$|b.\varphi(z).(z - a)^m| < |b.(z - a)^m|.$$

If we write $p_1(z) = b.(z - a)^m$, this means that

$$|z - a| = r \Rightarrow |p_1(z) - p(z)| < |p_1(z)|.$$

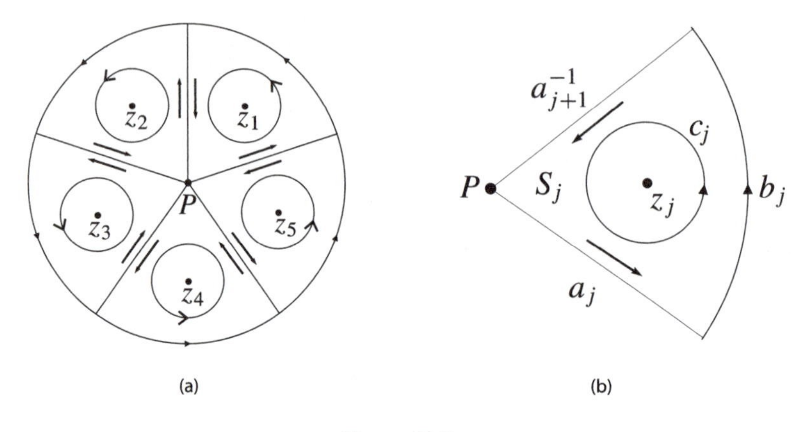

Figure 5.5.

By Rouché's theorem, it follows that $p(z)$ has, at the point $z = a$, the same local degree as $p_1(z)$, which equals m.

Proposition 5.3. *Let $p(z)$ be a polynomial that has no root z with $|z| = r$. Let $c_r\colon [0, 2\pi] \to \mathbb{C} - \{0\}$ be defined by $c_r(z) = r.e^{is}$. The winding number of the path $p \circ c_r$ around 0 equals the number of roots of $p(z)$ inside the disk $D = \{z \in \mathbb{C}; |z| \leq r\}$, each of these roots being counted according with its multiplicity.*

Proof. Let $z_1, \ldots, z_n$ be the roots of $p(z)$ that lie in the interior of D, each z_j having multiplicity m_j $(j = 1, \ldots, k)$. By drawing straight line segments from a certain point $P \in \text{int}. D$, we decompose the disk in a union of adjacent slices $S_1, \ldots, S_k$ (like a pizza pie) so that the root z_j lies in the interior or S_j, for $j = 1, \ldots, k$ (see Figure 5.5(a)). We treat the boundary of S_j as a path $\partial S_j = a_j.b_j a_{j+1}^{-1}$ if $1 \leq j \leq k-1$ and $\partial S_k = a_k.b_k a_1^{-1}$ (see Figure 5.5(b)). Here a_j is a linear path that starts at P and runs along the j-th line segment used to decompose D; b_j covers an arc of circle, so that $c_r \simeq b_1.b_2 \cdots b_k$. (See Corollary 2.1.)

For any differential form ω defined on an open set of the plane that contains D, we have

$$\int_{c_r} \omega = \sum_{j=1}^{k} \int_{b_j} \omega$$

and

$$\int_{\partial B_j} \omega = \int_{a_j} \omega + \int_{b_j} \omega - \int_{a_{j+1}} \omega \qquad (a_{j+1} = a_1 \quad \text{if} \quad j = k).$$

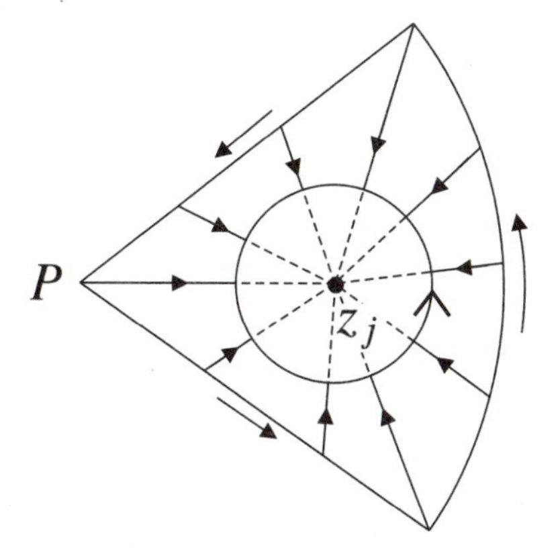

Figure 5.6.

Therefore,

$$\sum_{j=1}^{k} \int_{\partial B_j} \omega = \sum_{j=1}^{k} \int_{b_j} \omega = \int_{c_r} \omega.$$

Now if we let $c_j \colon [0, 2\pi] \to S_j$ be a path that parametrizes a small circle of center z_j in the usual manner, it is clear that ∂S_j is freely homotopic to c_j (by a linear radial homotopy with center z_j, as illustrated in Figure 5.6). Therefore, by item 2) of Proposition 3.6, $n(p \circ \partial S_j, 0) = n(p \circ c_j, 0)$, since $p(z) \neq 0$ at all points of S_j that are spanned by the homotopy.

Next, we take

$$\omega = \frac{ydx - xdy}{x^2 + y^2}.$$

The equality $n(p \circ \partial S_j, 0) = n(p \circ c_j, 0)$ means that

$$\frac{1}{2\pi} \int_{\partial S_j} p^*\omega = \frac{1}{2\pi} \int_{c_j} p^*\omega = m_j \qquad (= \text{multiplicity of the root } \; z_j).$$

Therefore,

$$n(c_r, 0) = \frac{1}{2\pi} \int_{c_r} p^*\omega = \sum_{j=1}^{k} \frac{1}{2\pi} \int_{c_j} p^*\omega = \sum_{j=1}^{k} m_j.$$

This completes the proof of the proposition. □

5.5 Exercises

1. Consider a continuous map $f\colon S^1 \to \mathbb{C} - \{0\}$. There exists $g\colon S^1 \to \mathbb{R}$ continuous such that $f(z) = |f(x)|e^{ig(z)}$ for every $z \in S^1$ if, and only if, the path $c\colon I \to \mathbb{C} - \{0\}$, defined by $c(s) = f(e^{2\pi is})$, satisfies $n(c, 0) = 0$.

2. Let $f\colon U \to \mathbb{C} - \{0\}$ be a continuous function in the open set $U \subset \mathbb{C}$. Prove that there exists a continuous function $g\colon U \to \mathbb{C}$ such that $f(z) = e^{g(z)}$ for every $z \in U$ if, and only if, for every closed path c in U, we have $n(f \circ c, 0) = 0$.

3. For every $n \in \mathbb{Z}$, draw a regular curve in the plane whose rotation number is n.

4. Show that the regular curve $c\colon [0, 2\pi] \to S^2$, given by $c(t) = (\cos t, \sin t, 0)$ is regularly homotopic in S^2 to its inverse c^{-1}.

5. Show that the figure eight is regularly homotopic in S^2 to the regular curve $c_2\colon [0, 2\pi] \to S^2$, $c_2(t) = (\cos 2t, \sin 2t, 0)$.

6. Prove that, with the notation of the two previous exercises, the regular curves c and c_2 are not regularly homotopic in S^2. (Hint: A regular curve in S^2, parametrized by arc length, is equivalent to a closed path in $\mathrm{SO}(3)$. A similar remark holds for regular homotopies.)

7. Show that there are precisely two regular homotopy classes of regular curves in S^2.

8. An *orthogonal couple* in $\mathbb{R}^3$ is an ordered pair (u, v) of non zero vectors such that $\langle u, v\rangle = 0$. Prove that there is no continuous field of orthogonal couples $(u, v)\colon D \to \mathbb{R}^3$ on the unit disc $D = \{(x, y, 0); x^2 + y^2 \leq 1\}$ with the property that $u(x, y, 0) = (x, y, 0)$, $v(x, y, 0) = (-y, x, 0)$ whenever $x^2 + y^2 = 1$.

9. Let $\omega = f dx + g dy$ be a differential form on an open convex set $U \subset \mathbb{R}^2$. Fix an arbitrary point $a \in U$ and define a function $F\colon U \to \mathbb{R}^2$ by taking, for every $z \in U$, $F(z) = \int_x \omega$, where $c\colon I \to U$ is given by $c(t) = (1-t)a + tz$. Prove that $dF = \omega$. As a consequence, prove that a differential form on an arbitrary open set $V \subset \mathbb{R}^2$ is closed if, and only if, it is locally exact. (That is, every point $a \in V$ is the center of a disc, restricted to which the form is exact.)

10. Let ω be a closed differential form on the open set $U \subset \mathbb{R}^2$. Given a piecewise C^1 path $c\colon J \to U$, with $J = [a, b]$, let $P = \{a = s_0 < s_1 < \cdots < s_n = b\}$ be a partition of J such that, for each $i = 1, \ldots, n$, the image

$c([s_{i-1}, s_i])$ is contained in a disc $D_i \subset U$. Let $g_i \colon D_i \to \mathbb{R}$ be such that $dg_i = \omega | D_i (i = 1, \ldots, n)$. Show that

$$\int_c \omega = \sum_{i=1}^{n} [g_i(c(s_i)) - g_i(c(s_{i-1}))].$$

11. In the context of the previous exercise, now let $c \colon J \to U$ be a continuous (but not necessarily piecewise C^1) path. Use the same construction as above to *define*

$$\int_c \omega = \sum_{i=1}^{n} [g_i(c(s_i)) - g_i(c(s_{i-1}))].$$

Prove that this sum neither depends on the partition P nor on the choices of the primitive functions g_i. (So the integral of a closed differential form can be defined along a path that is only continuous.)

12. Let $c, c' \colon J \to U$ be continuous paths with the same endpoints and such that, for each $s \in J$, the line segment $[c(s), c'(s)]$ is contained in U. Let ω be a closed differential form in U. Prove that $\int_c \omega = \int_{c'} \omega$. As a consequence, prove that if $c, c' \colon J \to U$ are two continuous paths such that $c \simeq c'$ and ω is a closed differential form in U, then $\int_c \omega = \int_{c'} \omega$.

13. Let $U \subset \mathbb{R}^2$ be simply connected. Prove that every closed differential form on U is exact.

14. Let $v \colon U \to \mathbb{R}^2$ be a continuous vector field. A *singularity* of v is a point $p \in U$ such that $v(p) = 0$. The *index* of the vector field v at an isolated singularity p is the local degree of v at p. Let $c \colon J \to U$ be a simple closed path whose interior contains a finite number of singularities of v. Prove that the winding number $n(v \circ c, O)$ equals the sum of the indices of v at the singularities that lie in the interior of c.

15. Let $c_1, c_2 \colon J \to \mathbb{R}^2 - \{0\}$ be closed paths of class C^1. If, for every $s \in J$, the tangent vectors $c_1'(s)$, $c_2'(s)$ are linearly independent, prove that $n(c_1, O) = n(c_2, O)$.

16. Let $\omega = f dx + g dy$ be a closed differential form on an open set $U \subset \mathbb{R}^2 - \{0\}$. Suppose that there exists a disc D with $O \in D \subset U$, and a constant k such that $z \in D \Rightarrow |f(z)| \leq k$ and $|g(z)| \leq k$. Prove that w is exact.

Part II
Covering Spaces

"Celui qui s'occupe beaucoup des mathématiques remarque, s'il a quelque expérience, que c'est une science très pauvre en pensées. Il n'y a en mathématiques pas plus d'idées primaires que de touches à un clavecin. Il n'est pas donné à un simple mortel d'augmenter à son gré le nombre de ces touches. Toute la joie d'un mathematicien c'est de jouer sur son clavecin. Le thème musical que nous voulons évoquer ici c'est la notion de recouvrement, et nous lui donnerons une extension assez générale pour qu'elle puisse servir de base à trois des plus belles théories mathématiques: a la théorie des fonctions de Riemann, au problème des formes spatiales et à la théorie des groupes continus".

W. Threlfall – La notion de recouvrement.
[L'Enseignement Mathématique vol. 34 (1935) pages 228-254.]

W. Threlfall co-authored, along with H. Seifert, two of the most beaufiful and inspiring topology books ever written: *Lehrbuch der Topologie* and *Variationsrechnung im Grossen.*

Chapter 6

Covering Spaces

6.1 Local Homeomorphisms and Liftings

Consider two topological spaces X, Y. A map $f\colon X \to Y$ is called a *local homeomorphism* if each point $x \in X$ is contained in an open set U such that $V = f(U)$ is open in Y and the restriction $f|U$ is a homeomorphism from U onto V.

Every (global) homeomorphism is, evidently, a local homeomorphism. A local homeomorphism $f\colon X \to Y$ is a continuous and open map. In particular, the image $f(X)$ is an open set in Y. It follows that if X is compact and Y is a connected Hausdorff space, every local homeomorphism $f\colon X \to Y$ is surjective. When we have a local homeomorphism $f\colon X \to Y$, the space X inherits all of the local topological properties from Y such as, for example, local connectivity, local compactness, and so on. If f is surjective, then Y also inherits the local topological properties of X. Given a local homeomorphism $f\colon X \to Y$ and an open subset $A \subset X$, the restriction $f|A$ is also a local homeomorphism from A onto $f(A)$. A surjective local homeomorphism $f\colon X \to Y$ is a quotient map; that is, $g\colon Y \to Z$ is continuous if, and only if, $g \circ f\colon X \to Z$ is continuous.

If $f\colon X \to Y$ is a local homeomorphism, then f is locally injective; that is, every point $x \in X$ has a neighborhood U such that $f|U$ is injective. But a continuous locally injective map, even when it is surjective, may not be a local homeomorphism. An example is given in Figure 6.1, where X is a segment, Y is a loop, and f is the obvious map from X onto Y. In this case, f is locally injective, we have $f(x_0) = f(x_1) = y_0$, but no neighborhood of x_0 (or of x_1) is transformed homeomorphically by f onto a neighborhood of y_0.

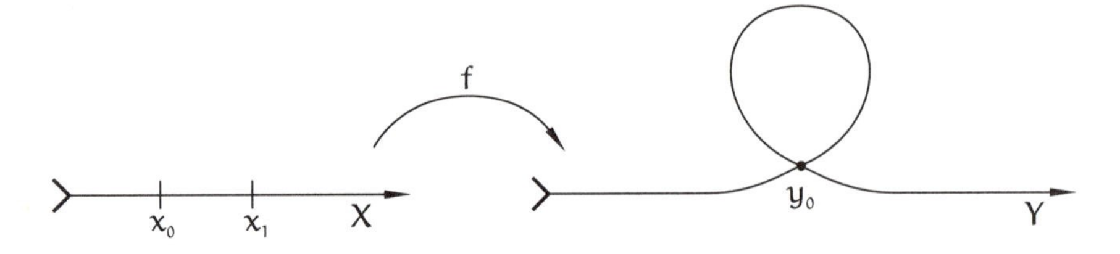

Figure 6.1. A locally injective map that is not a local homeomorphism.

Another example is given by $f\colon [0, 2\pi) \to S^1$, $f(t) = (\cos t, \sin t)$. Here, f is (globally) injective but it does not map any neighborhood of 0 (in the space $[0, 2\pi)$) onto a neighborhood of $f(0)$ in S^1.

A continuous locally injective map is a local homeomorphism if, and only if, it is open.

Example 6.1. The following maps are local homeomorphisms:

a. $\xi\colon \mathbb{R} \to S^1$, $\xi(t) = e^{it} = (\cos t, \sin t)$

b. $\zeta\colon \mathbb{R}^2 \to T = S^1 \times S^1$, $\zeta(s,t) = (e^{is}, e^{it})$

c. $f\colon \mathbb{R}^2 \to \mathbb{R}^2$, $f(x,y) = (e^x \cos y, e^x \sin y)$ or, using complex notation, $f(z) = e^z$

d. $\pi\colon S^n \to P^n$, $\pi(x) = \{x, -x\}$

e. $f\colon S^3 \to \mathrm{SO}(3)$, $f(x)(w) = x \cdot w \cdot x^{-1}$. (See Chapter 4, Section 1.)

With the exception of Case c, in all of the items above, the local homeomorphism is surjective. In Case c, we have $f(\mathbb{R}^2) = \mathbb{R}^2 - \{0\}$.

A more geometric version of Case a above can be obtained by considering an infinite spiral, say $X = \{(1 + e^t)e^{it}; t \in \mathbb{R}\}$, which turns around the unit circle S^1 (see Figure 6.2), and by defining $f\colon X \to S^1$ as the radial projection from the origin; that is, $f(z) = z/|z|$, or, $f((1 + e^t)e^{it}) = e^{it}$. ◁

In Case c, each vertical line passing through the point $(x, 0)$ is transformed by f onto a circle with center at the origin and radius e^x. The horizontal line through the point $(0, y)$ is transformed by f homeomorphically onto an open ray that starts at the origin and makes an angle of y radians with the positive x-axis. The inverse image $f^{-1}(b)$ of each point $b \in \mathbb{R}^2 - \{0\}$ is a countable set of points, all contained in the same vertical line, each one of them at a distance of 2π from the two other closest points. Every open horizontal strip with width 2π is transformed by f homeomorphically onto the complement of a ray ℓ which starts at the origin $0 = (0, 0)$.

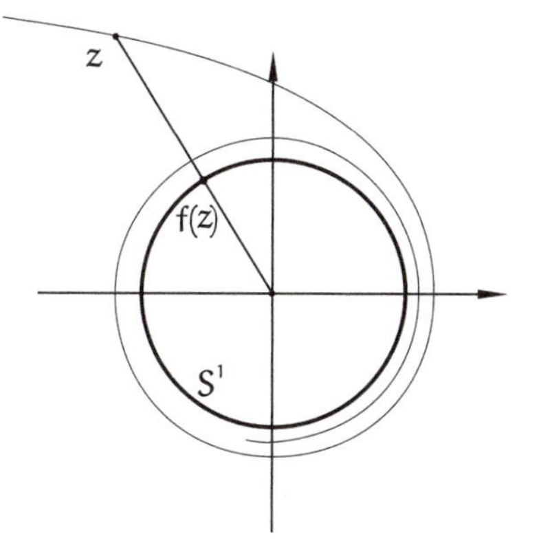

Figure 6.2. A local homeomorphism.

The inverse image $f^{-1}(\mathbb{R}^2 - \ell)$ is the countably infinite disjoint union of open horizontal strips of width 2π. Each one of these strips is transformed homeomorphically by f onto $\mathbb{R}^2 - \ell$.

Example 6.2. Let $U \subset \mathbb{R}^m$ be an open set and $f\colon U \to \mathbb{R}^m$ be a map of class C^1 whose derivative, $f'(x)\colon \mathbb{R}^m \to \mathbb{R}^m$, is an isomorphism at each point $x \in U$. The Inverse Function theorem guarantees that f is a local homeomorphism. Most examples of local homeomorphisms arise in this context, or in its global version, which can be stated as follows: *Let M^m, $N^m \subset \mathbb{R}^n$ be two differentiable manifolds, and f a map of class C^1 whose derivative $f'(x)\colon T_xM \to T_{f(x)}N$ is an isomorphism at each point $x \in M$. Then f is a local homeomorphism.* The five items in Example 6.1 are special cases of this situation. ◁

Proposition 6.1. *If the map $f\colon X \to Y$ is continuous and locally injective (in particular, a local homeomorphism), then the inverse image $f^{-1}(y)$ of each point $y \in Y$ is a discrete subset of X.*

Proof. Each point $x \in f^{-1}(y)$ has a neighborhood U, where x is the only point in U such that $f(x) = y$. Then $U \cap f^{-1}(y) = \{x\}$. So, every point $x \in f^{-1}(y)$ is isolated in $f^{-1}(y)$. □

Corollary 6.1. *Let X be a compact space, Y a Hausdorff space, and $f\colon X \to Y$ a continuous locally injective map. Then $f^{-1}(y)$ is finite for each $y \in Y$.*

Remark. Even if $f^{-1}(y)$ is finite for each $y \in Y$, the continuous map $f\colon X \to Y$ may not be locally injective. For example, the closed curve X,

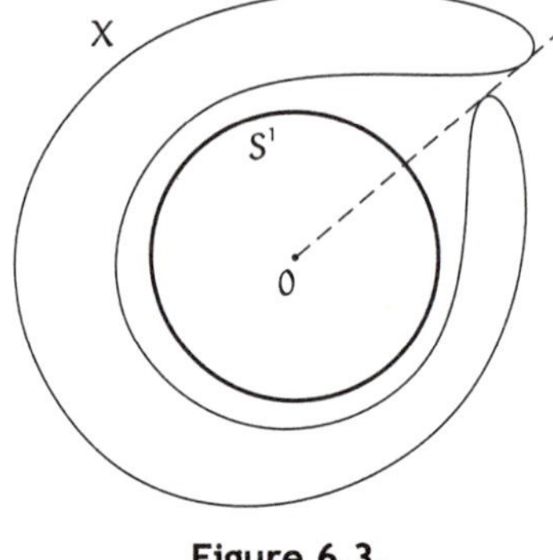

Figure 6.3.

sketched in Figure 6.3, projects radially onto the circle S^1 in such a way that the inverse image of each point of S^1 contains exactly two elements, but the radial projection of X onto S^1 is not locally injective.

Let $f\colon X \to Y$, $g\colon Z \to Y$ be two continuous maps. A *lifting* of g (with respect to f) is a continuous map $\widetilde{g}\colon Z \to X$ such that $f \circ \widetilde{g} = g$. This is illustrated in the diagram below.

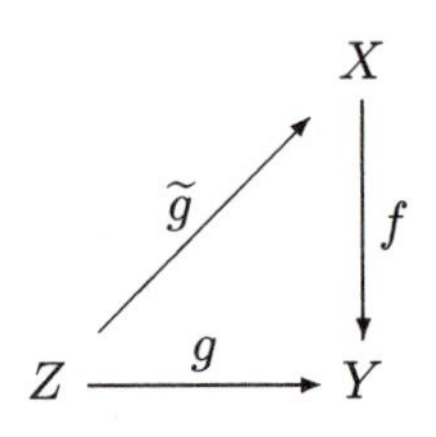

One of the basic problems that we study in this chapter is the existence and uniqueness of the lifting, in terms of the properties of the map f. We show now that if f is locally injective then the lifting of g, if it exists, is unique, provided that Z is connected, X is Hausdorff and we fix a value $\widetilde{g}(z_0)$. Note that not every continuous map g has a lifting, even when f is a local homeomorphism. (See Proposition 3.5.)

Proposition 6.2. *Let X be a Hausdorff space and $f\colon X \to Y$ be a continuous and locally injective map. If Z is connected and $g\colon Z \to Y$ is continuous, then two liftings $\widetilde{g}$, $\hat{g}\colon Z \to X$ of g, which coincide at one point $z \in Z$, are equal.*

Proof. The set $A = \{z \in Z, \widetilde{g}(z) = \hat{g}(z)\}$ is not empty, because $z_0 \in A$. Since X is a Hausdorff space, A is closed in Z. In order to conclude that $\widetilde{g} = \hat{g}$, we just have to prove that A is open in Z. For this, let $a \in A$. There exists a neighborhood V of $\widetilde{g}(a) = \hat{g}(a)$ such that $f|V$ is injective. By the continuity of $\widetilde{g}$ and $\hat{g}$, there exists a neighborhood U of a with $\widetilde{g}(U) \subset V$ and $\hat{g}(U) \subset V$. Hence, for all $z \in U$ we have $f\widetilde{g}(z) = g(z) = f\hat{g}(z)$ and, from the injectivity of f in V, $\widetilde{g}(z) = \hat{g}(z)$. Therefore $U \subset A$. □

Let $f: X \to Y$ be a continuous map. A *section* of f is a continuous map $\sigma: Y \to X$ such that $f \circ \sigma = id_Y$. In order to provide a section σ we must choose continuously, for each $y \in Y$, a point $\sigma(y)$ belonging to the inverse image (or fiber) $f^{-1}(y)$. This is not always possible. First of all, f must be surjective but this necessary condition is far from being sufficient, as we will see in what follows.

If $\sigma: Y \to X$ is a section of f then the restriction of f to $\sigma(Y)$ is a homeomorphism onto Y.

The following corollaries show some consequences of Proposition 6.2 with respect to the sections of a locally injective map.

Corollary 6.2. *Let X be a connected Hausdorff space. A continuous locally injective map $f: X \to Y$ that admits a section $\sigma: Y \to X$ is a homeomorphism and its inverse is σ.*

In fact, in this case the maps $\sigma \circ f$, $id_X: X \to X$ are liftings of f (relatively to f), as illustrated by the diagram below.

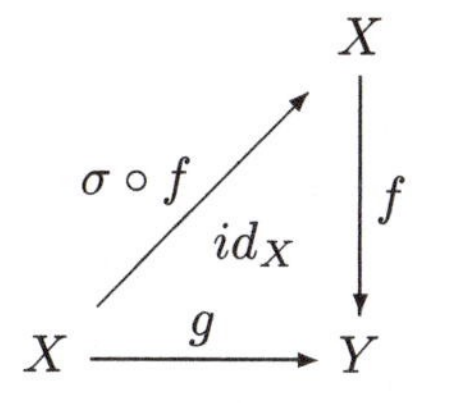

Since $\sigma \circ f$ coincides with id_X in the set $\sigma(Y)$, from Proposition 6.2 we conclude that $\sigma \circ f = id_X$, hence $\sigma = f^{-1}$.

It follows from Corollary 6.2 that a continuous, locally injective and non-injective map, whose domain is Hausdorff connected, does not admit a section. An example of such a map is $f: S^1 \to S^1$, $f(z) = z^2$.

Corollary 6.3. *Let X be a Hausdorff space, Y connected and $f: X \to Y$ a continuous, locally injective map. If $\sigma: Y \to X$ is a section of f then $\sigma(Y)$ is a connected component of X.*

In fact, let C be a connected component of X which contains the connected set $\sigma(Y)$. By Corollary 6.2, $f|C$ is a homeomorphism from C onto Y. Since $f|\sigma(Y)$ is already a homeomorphism onto Y, we have $\sigma(Y) = C$.

Corollary 6.4. *Let A, B be open and connected subsets in the Hausdorff space $X = A \cup B$ and $f: X \to Y$ a continuous map such that $f|A$ and $f|B$ are homeomorphisms onto Y. Then $A \cap B = \varnothing$ or $A = B$.*

In fact, f is locally injective, $Y = f(A)$ is connected and $(f|A)^{-1}: Y \to X$ is a section. By Corollary 6.3, $A = (f|A)^{-1}(Y)$ is a connected component of X. In a similar way we show that B is also a connected component. It follows that $A = B$ or $A \cap B = \varnothing$.

Remark. Corollary 6.4 would be false without the hypothesis that A and B are open sets. This is shown by the function $f: S^1 \to [-1, 1]$, defined by $f(x, y) = x$, and the sets $A = \{(x, y) \in S^1; y \geq 0\}$, $B = \{(x, y) \in S^1; y \leq 0\}$.

6.2 Covering Maps

The Inverse Function Theorem is usually employed to prove that a certain map $f: X \to Y$ is a local homeomorphism, but a natural question remains open: Is f a (global) homeomorphism from X onto $f(X)$? Since f is already an open map, this is equivalent to ask if the map f is injective. This is a global question, of topological nature, whose answer cannot be given by Differential Calculus theorems, which are essentially local. We will discuss this problem here.

A local homeomorphism $f: X \to Y$ can be interpreted from the following viewpoint: given $a \in X$ and $b = f(a) \in Y$, the equation $f(x) = y$ has, for each y sufficiently close to b, a unique solution x, close to a, which depends continuously on y. It remains to be known the conditions under which this locally unique solution is globally unique in X.

The classical (and the most adequate) instrument to investigate if a given local homeomorphism is global and, more generally, to obtain regions where the homeomorphism is injective, is the *method of analytic continuation*, which we briefly describe now. Given a local homeomorphism $f: X \to Y$, let $y \in f(X)$. For each $x \in X$ with $f(x) = y$, there exist neighborhoods $U \ni x$ and $V \ni y$ such that $f: U \to V$ is a homeomorphism; $g = (f|U)^{-1}: V \to U$ is a local inverse of f, known classically as a "branch of f^{-1}." The problem consists of extending this branch g to a region larger than V. Given another point $y' \in f(X)$, we connect y' to y by a path a and we try to extend g along this path. This is possible provided that the path a has a lifting; that is, there exists a path $\widetilde{a}$ in X, with initial point x, such that $f(\widetilde{a}(s)) = a(s)$ for all $s \in I$. Then, since $y' = a(1)$, we define $g(y') = \widetilde{a}(1)$. The existence of the lifting $\widetilde{a}$ cannot be guaranteed, as the example below shows. The concept of *covering*, which we introduce below,

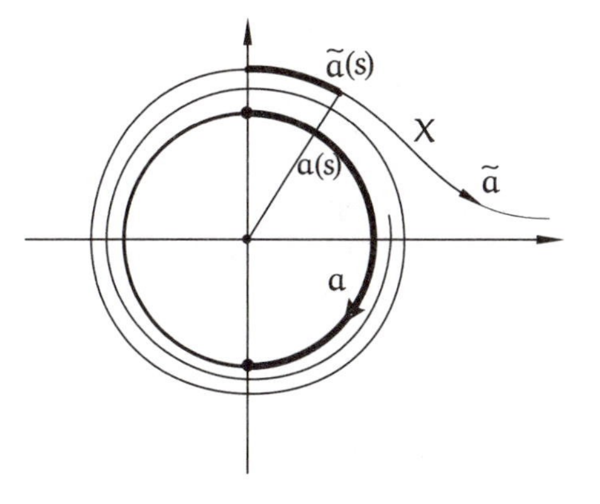

Figure 6.4.

provides additional conditions that will enable the use of the method of analytical continuation.

Example 6.3. Let $f\colon (0,3\pi)\to S^1$ be the surjective local homeomorphism defined by $f(t)=(\cos t,\sin t)$. It is easy to obtain, in the circle S^1, paths that cannot be lifted relatively to f. We just have to take, for example, a closed path in S^1 whose degree is ≥ 2. For another example, consider the set $X=\{(1+t)e^{i/t}; 0<t<+\infty\}$ and define $g\colon X\to S^1$ as the radial projection, $g(z)=z/|z|$. By taking $t=2/\pi$, we see that the point $\left(0,1+\frac{2}{\pi}\right)$ belongs to X. Let $a\colon I\to S^1$ be a path with origin $(0,1)$, which describes homeomorphically the semicircle $x>0$, and ends at the point $(0,-1)$ (see Figure 6.4). There is no path $\widetilde{a}\colon I\to X$ with $\widetilde{a}(0)=\left(0,1+\frac{2}{\pi}\right)\in X$ such that $g\circ\widetilde{a}=a$, even though we have $g(\widetilde{a}(0))=a(0)$. ◁

A map $p\colon \widetilde{X}\to X$ is called a *covering map* (or, simply, a *covering*) when each point $x\in X$ belongs to an open set $V\subset X$ such that

$$p^{-1}(V)=\bigcup_{\alpha}U_\alpha$$

is a union of pairwise disjoint open sets U_α such that, for each α, the restriction $p|U_\alpha\colon U_\alpha\to V$ is a homeomorphism. The open set V satisfying the above condition is called a *distinguished neighborhood.* The space $\widetilde{X}$ is called a *covering space* of X and, for each $x\in X$, the set $p^{-1}(x)$ is called a *fiber* over x. Sometimes, X is called the *base* of the covering.

A covering map $p\colon \widetilde{X}\to X$ is a local homeomorphism from $\widetilde{X}$ onto X. Example 6.4 shows that not every local homeomorphism is a covering map.

The local homeomorphisms in Example 6.1 are covering maps. When the space Y is discrete, the projection $p\colon X \times Y \to X$ is a covering map.

Every open subset of a distinguished neighborhood is itself a distinguished neighborhood. Thus, when X is locally connected, locally compact, etc., we may choose the distinguished neighborhoods in such a way to be connected, with compact support, and so on.

If X is a locally connected and locally Hausdorff space, each distinguished neighborhood V can be chosen to be connected and Hausdorff. Thus we do not need to suppose that, in the decomposition

$$p^{-1}(V) = \bigcup_{\alpha} U_\alpha,$$

where $p|U_\alpha$ is, for each α, a homeomorphism onto V, the open sets U_α be pairwise disjoint. (See Corollary 6.4.) In this case, the sets U_α are the connected components of the inverse image $p^{-1}(V)$.

When $p\colon \widetilde{X} \to X$ is a covering, the condition that $\widetilde{X}$ be a Hausdorff space can be omitted from Proposition 6.2. We have the following proposition.

Proposition 6.3. *Let $p\colon \widetilde{X} \to X$ be a covering map and Z a connected space. If $\widetilde{g}$, $\hat{g}\colon Z \to \widetilde{X}$ satisfy $p \circ \widetilde{g} = p \circ \hat{g} = g$, then either $\widetilde{g}(z) \neq \hat{g}(z)$ for all $z \in Z$ or $\widetilde{g} = \hat{g}$.*

Proof. Since p is locally injective, we use the proof of Proposition 6.2 in order to see that the set $A = \{z \in Z; \widetilde{g}(z) = \hat{g}(z)\}$ is open. In order to show that A is closed, without using that $\widetilde{X}$ is a Hausdorff space, choose $z \in Z$ such that $\widetilde{g}(z) \neq \hat{g}(z)$. The image of these two points by the map p is the same point $g(z) \in X$. Let V be a distinguished neighborhood of $g(z)$. Then

$$p^{-1}(V) = \bigcup_{\alpha} U_\alpha,$$

the disjoint union of open set which are mapped homeomorphically by p onto V. Therefore, there exists $\alpha \neq \beta$ such that $\widetilde{g}(z) \in U_\alpha$ and $\hat{g}(z) \in U_\beta$. By taking an open neighborhood $W \ni z$ in Z such that $\widetilde{g}(W) \subset U_\alpha$ and $\hat{g}(W) \subset U_\beta$, we see that $\widetilde{g}(w) \neq \hat{g}(w)$ for all $w \in W$. Hence, $z \notin A \Rightarrow z \in W$ with $W \cap A = \varnothing$. Thus, A is closed. □

Proposition 6.4. *If the base X of a covering $p\colon \widetilde{X} \to X$ is connected, then each fiber $p^{-1}(x)$ $x \in X$, has the same cardinal number, which is called the* number of leaves *of the covering.*

Proof. For every point x of a distinguished neighborhood V, the cardinal number of the fiber $p^{-1}(x)$ is the same. Hence, the set of the points $x \in X$ such that $p^{-1}(x)$ has a prescribed cardinal number is open. This determines a decomposition of X as the union of disjoint open sets, where in each of them the cardinal number of $p^{-1}(x)$ is constant. Since X is connected, this family of disjoint open sets has only one set. □

Remarks. 1. By Corollary 6.1, when $\widetilde{X}$ is compact and X is a connected Hausdorff space, every covering map $p\colon \widetilde{X} \to X$ has a finite number of leaves. In this case, $X = p(\widetilde{X})$ is necessarily compact.

2. A covering map $p\colon \widetilde{X} \to X$ whose base X is connected is a locally trivial fibration whose typical fiber F is discrete (and whose cardinality is equal to the number of leaves of p).

Example 6.4. Let $f\colon X \to Y$ be a local homeomorphism, where Y is connected and each inverse image $f^{-1}(y)$, $y \in Y$, is finite. Given any $x_0 \in X$, the restriction $f_0 = f|(X - \{x_0\})$ is still a local homeomorphism. But at least one of the maps, f or f_0, is not a covering map. In fact, either the number of elements of $f^{-1}(y)$ is not constant or the number of elements of $f_0^{-1}(y)$ is not constant. ◁

It is important to recognize when a local homeomorphism $f\colon X \to Y$ is a covering map. Now we characterize the coverings with a finite number of leaves. Other sufficient conditions will be studied later on.

A map $f\colon X \to Y$ is called *closed* when the image $f(F)$ of every closed subset $F \subset X$ is a closed subset of Y.

In order that a map $f\colon X \to Y$ be closed, it is necessary and sufficient that, for every $y \in Y$ and every open set $U \supset f^{-1}(y)$ in X, there exists an open set $V \ni y$ in Y such that $f^{-1}(V) \subset U$. (See the Appendix at the end of the book.) Note that this condition suggests something like the continuity of the correspondence $y \mapsto f^{-1}(y)$, which is not a function.

A continuous map $f\colon X \to Y$ is called *proper* when it is closed and, for every $y \in Y$, the inverse image $f^{-1}(y)$ is compact. In the Appendix, we prove some basic properties of proper maps.

Proposition 6.5. *Let X be a Hausdorff space and $f\colon X \to Y$ a local homeomorphism. Each of the following statements implies the next one:*

1. *There exists $n \in \mathbb{N}$ such that each inverse image $f^{-1}(y)$, $y \in Y$, has n elements.*

2. *f is proper and surjective.*

3. f is a covering map whose fibers $f^{-1}(y)$ are finite.

If Y is connected, then the three statements are equivalent.

Proof. $1 \Rightarrow 2$. We just need to prove that f is closed. Let $y \in Y$ and $A \supset f^{-1}(y)$ be an open set. Since $f^{-1}(y) = \{x_1, \ldots, x_n\}$ is finite and X is Hausdorff, there exist pairwise disjoint open sets $W_1 \ni x_1, \ldots, W_n \ni x_n$, such that $W_1 \cup \ldots \cup W_n \subset A$. Hence,

$$V = \bigcap_{i=1}^{n} f(W_i)$$

is an open neighborhood of y. For each $i = 1, \ldots, n$, $U_i = W_i \cap f^{-1}(V)$ is open and, by setting $U = \cup U_i$, we have $f^{-1}(V) \supset U$. We claim that we must have $f^{-1}(V) = U$. In fact, if $w \in f^{-1}(V)$—that is, $f(w) = v \in V = \cap f(W_i)$—then there exist $w_1 \in W_1, \ldots, w_n \in W_n$ such that $f(w_i) = v$ for every i. Since $f^{-1}(v)$ has n elements and the sets W_i are pairwise disjoint, we must have $w = w_i$ for some i, thus $w \in U_i = f^{-1}(V) \cap W_i$, that is, $w \in U$. Hence, $f^{-1}(V) = U \subset \cup W_i \subset A$, which proves that f is closed, because of the criterion about closed maps mentioned above and proved in the Appendix.

$2 \Rightarrow 3$. Given an arbitrary point $y \in Y$, its inverse image is a compact discrete set, therefore, it is finite: $f^{-1}(y) = \{x_1, \ldots, x_n\}$. Let $W_1 \ni x_1, \ldots W_n \ni x_n$ be pairwise disjoint open sets in X, which are mapped homeomorphically by f onto open sets of Y. Then $f(W_1) \cap \ldots \cap f(W_n)$ is an open neighborhood of y and, since f is closed, we can obtain an open set V with $y \in V \subset \cap f(W_i)$ and such that $f^{-1}(V) \subset \cup W_i$. For every $i = 1, \ldots, n$, we take $U_i = f^{-1}(V) \cap W_i$. Then $f^{-1}(V) = (\cup W_i) \cap f^{-1}(V) = \cup(f^{-1}(V) \cap W_i) = \cup U_i$ and, since $V \subset f(W_i)$, f maps each one of the open sets U_i homeomorphically onto V.

Finally, when Y is connected, $3 \Rightarrow 1$ by Proposition 6.4. □

Corollary 6.5. *If X is a Hausdorff compact space and Y is Hausdorff, then every surjective local homeomorphism $f\colon X \to Y$ is a covering.*

In fact, f is proper.

Corollary 6.6. *Let X, Y be Hausdorff spaces. If $A \subset X$ has compact closure and the local homeomorphism $f\colon A \to Y$ extends continuously to a map $\bar{f}\colon \overline{A} \to Y$ such that $\bar{f}(\partial A) \subset \partial \bar{f}(A)$ (that is, $\bar{f}$ maps the boundary of A into the boundary of $\bar{f}(A)$), then the restriction $f|A\colon A \to f(A)$ is a covering.*

In fact, under these conditions, f is a proper map from A onto $f(A)$.

Remarks. 1. Let $f\colon X \to Y$ be a covering map. If Y is a Hausdorff compact space and each fiber $f^{-1}(y)$ is finite, then X is compact. The proof is easy (even in the general case where f is a locally trivial fibration, with compact base and fiber). By supposing that X is Hausdorff and Y is connected, it follows as a corollary of Proposition 6.5, because f is proper and $X = f^{-1}(Y)$.

2. In order to prove that $1 \Rightarrow 2$ we used only the fact that f is open and continuous. Therefore, we can state that if $f\colon X \to Y$ is a continuous open map, and there exists $n \in \mathbb{N}$ such that all of the inverse images $f^{-1}(y)$, $y \in Y$ have n elements, then f is a covering map. About the necessity that f be an open map, see Figure 6.3.

3. In the Appendix, Proposition A.4, we show that if X and Y are metric spaces without isolated points, a local homeomorphism $f\colon X \to Y$ which is also a closed map is necessarily a proper map.

Example 6.5. The maps in the items a), b), and c) in Example 6.1 are not proper maps; the maps in items d) and e) are proper. The map $f\colon S^1 \to S^1$, defined by $f(z) = z^n$, is a local homeomorphism (by the Inverse Function theorem). Since S^1 is a Hausdorff compact space and f is surjective, we see that it is a covering map with n leaves. $\triangleleft$

Example 6.6. Let $p\colon \mathbb{C} \to \mathbb{C}$ be a non constant complex polynomial and $F \subset \mathbb{C}$ the finite set whose elements are the roots of $p'(z)$. By setting $X = \mathbb{C} - p^{-1}(p(F))$ and $Y = \mathbb{C} - p(F)$, we see that the restriction $p|X\colon X \to Y$ is a local homeomorphism and a proper map; $p|X$ is surjective because Y is connected. Hence, $p|X$ is a covering with n leaves, where n is the degree of p. In particular, the map $p\colon \mathbb{C} - \{0\} \to \mathbb{C} - \{0\}$, $p(z) = z^n$, is a covering. $\triangleleft$

6.3 Properly Discontinuous Groups

Important examples of covering maps are obtained when we consider properly discontinuous groups of homeomorphisms, which we study now.

The set of homeomorphisms of a topological space X is a group with the operation of composition. A subgroup G of this group is called a *group of homeomorphisms* of X. Therefore, we must have:

1. $id_X \in G$;
2. $g, h \in G \Rightarrow gh \in G$;
3. $g^{-1} \in G$.

(Here, gh is the composition of g with h.)

For the sake of simplicity in the notation, the image of the point x by the homeomorphism $g\colon X \to X$ is denoted by gx.

The *orbit* of a point $x \in X$ relative to a group of homeomorphisms G is the set $G \cdot x = \{gx; g \in G\}$. The relation "there exists $g \in G$ such that $gx = y$" is an equivalence relation on the set X. The equivalence class of a point $x \in X$ according to this relation is the orbit of the point $G \cdot x$. Therefore, given x, $y \in X$, either $G \cdot x = G \cdot y$ or $G \cdot x \cap G \cdot y = \varnothing$.

A group G of homeomorphisms of a space X is said to be *properly discontinuous* when every point $x \in X$ has a neighborhood V such that, for every $g \in G$ different from the identity, we have $g \cdot V \cap V = \varnothing$. Equivalently: If $g \neq h$ in G, then $g \cdot V \cap h \cdot V = \varnothing$. We say that V is a *convenient neighborhood* of the point x.

If G is a properly discontinuous group of homeomorphisms of a topological space X, then for every $g \neq id_X$ in G and every $x \in X$, we have $gx \neq x$. That is, with the exception of the identity, the homeomorphisms that belong to G do not have fixed points. This is equivalent to stating that $g \neq h$ in $G \Rightarrow gx \neq hx$ for all $x \in X$. We also say, in this case, that G operates *freely* in X.

Given a properly discontinuous group G of homeomorphisms of a space X, the orbit $G \cdot x$ of each point of X is a discrete set. In fact, if V is a convenient neighborhood of the point x then each set $g \cdot V$, $g \in G$ is a neighborhood of gx which contains only this point of the orbit $G \cdot x$.

A neighborhood V is convenient with respect to a properly discontinuous group G if, and only if, it contains at most one element of any orbit of G. In fact, let V be a convenient neighborhood. If there exist y, $gy \in V$ then $gy \in V \cap gV$; hence, $g = id_X$ and from this, $gy = y$. Conversely, if V does not contain two distinct elements of any orbit of G then, for all $y \in V$ and $g \neq id_X$ in G, we have $gy \notin V$; that is, $V \cap g \cdot V = \varnothing$.

If the points of the space X are closed sets (for example, if X is a Hausdorff, or locally Hausdorff, space), then the orbits $G \cdot x$ relative to a properly discontinuous group G of homeomorphisms of X are closed subsets of X. In fact, if $y \notin G \cdot x$, then a convenient neighborhood of y contains, at most, one point gx of the orbit $G \cdot x$. Since $\{gx\}$ is a closed set, a smaller neighborhood of y will be disjoint of $G \cdot x$.

In particular, when X is a compact Hausdorff space, every properly discontinuous group G of homeomorphisms of X is finite. In fact, by fixing $x_0 \in X$, the map $g \mapsto gx_0$ is a bijection from G onto the orbit $G \cdot x_0$ which, being a discrete and closed subset of the compact space X, is finite.

The condition that the points of X be closed is essential in order that the orbits of a properly discontinuous group of homeomorphisms of X be closed. This can be easily seen by considering the group $G = \{id_X\}$.

Example 6.7. For each $m \in \mathbb{Z}$, let $T_m \colon \mathbb{R} \to \mathbb{R}$ be the translation $T_m(x) = x + m$. The set $G = \{T_m; m \in \mathbb{Z}\}$ is a properly discontinuous group of homeomorphisms of $\mathbb{R}$. More generally, let $\mathbb{Z}^n \subset \mathbb{R}^n$ be the additive subgroup that consists of the vectors whose coordinates are integers. For each $v \in \mathbb{Z}^n$, let $T_v \colon \mathbb{R}^n \to \mathbb{R}^n$ be the translation $T_v(x) = x + v$. The set $G = \{T_v; v \in \mathbb{Z}^n\}$ is a properly discontinuous group of homeomorphisms of $\mathbb{R}^n$. Any neighborhood whose diameter is smaller than 1 is a convenient neighborhood for this group. $\triangleleft$

Example 6.8. Let $\alpha \colon S^m \to S^m$ be the antipodal map. The set $G = \{id, \alpha\}$ is a group of homeomorphisms of S^m since $\alpha \circ \alpha = id$. G is properly discontinuous because if V is an open set contained in a hemisphere, then $\alpha \cdot V \cap V = \varnothing$. $\triangleleft$

Example 6.9. Let G be a finite group of homeomorphisms of a Hausdorff space X such that, with the exception of the identity, no element $g \in G$ has fixed points. Then G is properly discontinuous. In fact, given $x \in X$, if $g \neq h$ in G, we have $gx \neq hx$. By Hausdorff axiom it is possible to obtain, for each $g \in G$, an open set V_g containing gx, such that $g \neq h$ implies that $V_g \cap V_h = \varnothing$. By the continuity of the homeomorphisms $g \in G$ and the fact that G is finite, we can take a neighborhood $V = V_{id}$ of x so small that $g \cdot V \subset V_g$ for every $g \in G$. Then $g \cdot V \cap V = \varnothing$ for every $g \in G$. Note that Example 6.8 is a particular case of this one. $\triangleleft$

Example 6.10. Let G be a topological group. For each subgroup $H \subset G$ we may consider the group $\ell(H)$ of homeomorphisms of G, whose elements are the left translations $\ell_h \colon G \to G$, $\ell_h(x) = h \cdot x$, defined by elements $h \in H$. (See Example 6.7 where $G = \mathbb{R}^n$ e $H = \mathbb{Z}^n$.) The group of homeomorphisms $\ell(H)$ is properly discontinuous if, and only if, H is a discrete subgroup of G. One part of the statement is obvious: if $\ell(H)$ is properly discontinuous, the orbit of each element of G is a discrete set. In particular, H is discrete because it is the orbit of the neutral element of G. Conversely, suppose that $H \subset G$ is discrete. Then there exists a neighborhood U of the neutral element $e \in G$ such that $U \cap H = \{e\}$. Since the map $(x, y) \mapsto xy^{-1}$, of $G \times G$ into G, is continuous, there exists a neighborhood $V \ni e$ such that $x, y \in V \Rightarrow xy^{-1} \in U$. We assert that, for every $h \in H$, with $h \neq e$, we have $(\ell_h \cdot V) \cap V = \varnothing$. In fact, if there existed $x \in (\ell_h \cdot V) \cap V$ then there would exist $y \in V$ with $x = hy$ and from this $h = xy^{-1} \in U \cap H$, hence $h = e$. Note that the orbits of the group of homeomorphisms $\ell(H)$ are the cosets $H \cdot x$, determined by the subgroup $H \subset G$. Example 6.7 is a particular case of this situation. $\triangleleft$

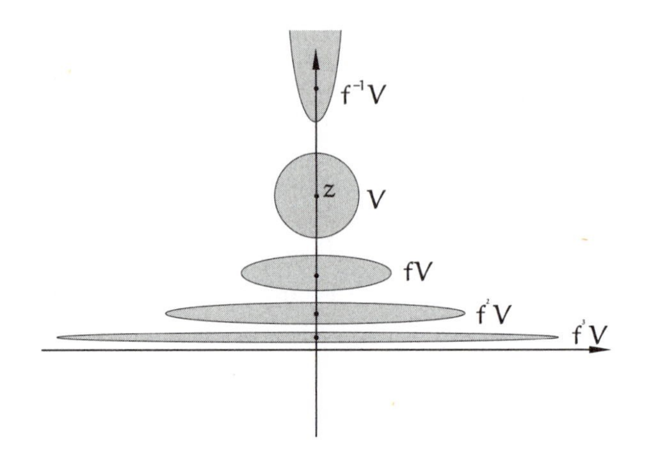

Figure 6.5.

Example 6.11. Let $X = \mathbb{R}^2 - \{0\}$. The homeomorphism $f\colon X \to X$, defined by $f(x, y) = (ax, a^{-1}y)$, where $a > 1$ is a constant, generates a group G of homeomorphisms of X, whose elements are the powers f^n, $n \in \mathbb{Z}$, where $f^n = f \circ f \circ \cdots \circ f$ (n times) if $n > 0$, $f^n = f^{-1} \circ \ldots \circ f^{-1}$ ($|n|$) times) if $n < 0$ and $f^0 = id_X$. The group G is properly discontinuous. In fact, for every $n \in \mathbb{Z}$, we have $f^n(x, y) = (a^n \cdot x, a^{-n} \cdot y)$. If G were not properly discontinuous, there would exist a point $z = (x, y) \in X$, a sequence of points $z_n = (x_n, y_n) \in X$ and a sequence of integers $k_n \neq 0$ such that $\lim_n z_n = z$ and $\lim_n f^{k_n}(z_n) = z$. In order to fix ideas, suppose that $x \neq 0$. Then, from the hypothesis $\lim(x_n, y_n) = (x, y) = \lim(a^{k_n} \cdot x_n, a^{-k_n} \cdot y_n)$, it would result that $\lim_n a^{k_n} = 1$, which contradicts $k_n \in \mathbb{Z} - \{0\}$.

Figure 6.5 shows the orbit of a point $z = (0, y)$, whose elements are the points $z_n = (0, a^{-n} \cdot y)$, and the sets $f^n V$, $n \in \mathbb{Z}$, where V is a disk with center z. ◁

Given a group G of homeomorphisms of X, we denote by X/G the quotient space of X by the equivalence relation whose equivalence classes are the orbits Gx, $x \in X$. The canonical projection $p\colon X \to X/G$ associates to each point $x \in X$ its orbit $p(x) = G \cdot x$. The open sets of the topology of X/G are the sets $A \subset X/G$ such that $p^{-1}(A)$ is open in X. Thus, the open sets of X/G are the images $p(U)$ where $U \subset X$ is an open set which is a union of orbits.

The continuous map $p\colon X \to X/G$ is open because if $V \subset X$ is open, then $p^{-1}(p(V)) = \bigcup_{g\in G} g{\cdot}V$ is open in X.

Proposition 6.6. *Let G be a group of homeomorphisms freely operating in the space X. The following statements are equivalent:*

1. *G is properly discontinuous.*
2. *The canonical projection $p\colon X \to X/G$ is a covering map.*
3. *$p\colon X \to X/G$ is locally injective.*

Proof. $1 \Rightarrow 2$: Let $y = p(x)$ be an arbitrary point in X/G. Take a convenient neighborhood $U \ni x$. Since p is an open map, $V = p(U)$ is an open neighborhood of y. We have that

$$p^{-1}(V) = \bigcup_{g\in G} g{\cdot}U$$

is the union of pairwise disjoint open sets (because U is a convenient neighborhood), and the restriction of the continuous map p to each of these open sets is injective; therefore, it is a homeomorphism onto $p(g{\cdot}U) = p(U) = V$. Hence 1 implies 2.
$2 \Rightarrow 3$: Obvious.
$3 \Rightarrow 1$: From 3, we conclude that each point $x \in X$ belongs to an open set U in which there are no two points in the same orbit. Then U is a convenient neighborhood of x and this proves that $3 \Rightarrow 1$. □

Corollary 6.7. *Let $f\colon G \to H$ be a surjective continuous homomorphism between two topological groups. In order that f be a covering map, it is necessary and sufficient that it be a local homeomorphism or, equivalently, that f be continuous, open, and its kernel be a discrete subgroup.*

In fact, under these conditions, denoting by $K = f^{-1}(e)$ the kernel of f, the group $\ell(K)$ of left translations by elements of K operates in a properly discontinuous mode in G (see Example 6.10). Hence, the quotient map $\pi\colon G \to G/K$ is a covering. By passing to the quotient, there exists a homeomorphism $\bar{f}\colon G/K \to H$ such that $\bar{f}\circ\pi = f$. Hence, f is a covering. The converse is obvious.

The quotient space X/G of a Hausdorff space X by a properly discontinuous group of homeomorphisms G is locally Hausdorff because it is locally homeomorphic to X. But, globally, X/G may or may not be a Hausdorff space. Since $p\colon X \to X/G$ is open, the necessary and sufficient condition in order that X/G be Hausdorff is that the set $\Gamma = \{(x, gx); x \in X, g \in G\}$, graph of the equivalence relation determined by G, be closed in $X \times X$. When G is finite, then Γ is the union of a finite number of closed subsets of $X \times X$ (the graphs of the homeomorphisms $g \in G$). Hence, X/G is Hausdorff. In particular, when the Hausdorff space X is compact, the quotient space X/G of X by a properly discontinuous group of homeomorphisms is Hausdorff because G is necessarily finite.

In Example 6.11 above, the quotient space X/G is not Hausdorff. In fact, the points $w = (0, 1)$ and $z = (1, 0)$ do not belong to the same orbit. Nevertheless, for any disks $U \ni w$ and $V \ni z$, we have that $f^n U$ is, for large values of $n > 0$, a long flattened oval, close to the x-axis. (On the other hand, for large $n < 0$, $f^n U$ is a long vertical oval, close to the y-axis.) This forces $f^n U \cap V \neq \varnothing$ for $n > 0$ sufficiently large. From this, it follows that the points $G \cdot w$ and $G \cdot z$ in X/G do not have disjoint neighborhoods.

Another way to verify that X/G is not a Hausdorff space is to consider the sequences $w_n = (a^{-n}, 1)$ and $z_n = (1, a^{-n})$. For each $n \in \mathbb{N}$, w_n and z_n belong to the same orbit because $z_n = f^n w_n$. But $\lim w_n = (0, 1) = w$ and $\lim z_n = (1, 0) = z$ belong to distinct orbits of G. This says that, in the quotient space X/G, the sequence $p(w_n) = p(z_n)$ has two distinct limits $p(w)$ and $p(z)$. Hence, X/G is not Hausdorff. The reader may imagine the quotient space X/G as the union of four cylinders and four circles in $\mathbb{R}^3$, with a topology different from the usual.

In Example 6.7, the quotient space $\mathbb{R}^n/G = \mathbb{R}^n/\mathbb{Z}^n$ is the n-dimensional torus. For $n = 1$, we obtain the circle S^1 and, in general, $\mathbb{R}^n/\mathbb{Z}^n$is homeomorphic to the Cartesian product $S^1 \times \ldots \times S^1$ of n copies of the circle. Thus, even with G being infinite, the quotient space is Hausdorff.

In Example 6.8, the quotient space S^n/G is the n-dimensional projective space.

6.4 Path Lifting and Homotopies

A continuous and surjective map $f\colon X \to Y$ is said to have the *path lifting property* when, for any arbitrary path $a\colon J \to Y$, with $J = [s_0, s_1]$, and each point $x \in X$ such that $f(x) = a(s_0)$, there exists a path $\widetilde{a}\colon J \to X$ such that $\widetilde{a}(s_0) = x$ and $f \circ \widetilde{a} = a$.

We know that not every local homeomorphism $f\colon X \to Y$ has the path lifting property but, when X is a Hausdorff space, the lifting $\widetilde{a}\colon J \to X$ of a path $a\colon J \to Y$ is completely determined by a and the initial point $x = \widetilde{a}(s_0)$.

When f is surjective and, for any arbitrary path $a\colon J \to Y$ and any point $x \in X$ with $f(x) = a(s_0)$, there exists a unique path $\widetilde{a}\colon J \to X$ such that $f \circ \widetilde{a} = a$ and $\widetilde{a}(s_0) = x$, we say that $f\colon X \to Y$ has the *unique path lifting property*.

Even when $\widetilde{X}$ is not Hausdorff, a covering $p\colon \widetilde{X} \to X$ has the unique path lifting property. This is the content of the proposition below, according to which the analytic continuation of the local inverse of p along a path is always possible when the local homeomorphism p is a covering map. The reader should not forget that the unique path lifting property requires, first of all, that f be surjective, by definition.

Proposition 6.7. *Let $p\colon \widetilde{X} \to X$ be a covering map. Given a path $a\colon J \to X$, $J = [s_0, s_1]$ and a point $\widetilde{x} \in \widetilde{X}$ such that $p(\widetilde{x}) = a(s_0)$, there exists a unique path $\widetilde{a}\colon J \to \widetilde{X}$ with $\widetilde{a}(s_0) = \widetilde{x}$ and $p \circ \widetilde{a} = a$. (In other words: p has the unique path lifting property.)*

Proof. Assume initially that $a(J) \subset V$, where V is a distinguished neighborhood. Then, since $\widetilde{x} \in p^{-1}(V)$, there exists an open set $U \ni \widetilde{x}$ which is mapped homeomorphically by p onto V. Let $f = (p|U)^{-1}\colon V \to U$, and set $\widetilde{a} = f \circ a$. Next, consider the case where $J = J_1 \cup J_2$ is the union of two compact intervals with an endpoint s_* in common, in such a way that the proposition holds for the restrictions $a_1 = a|J_1$ and $a_2 = a|J_2$. We choose $\widetilde{a}_1\colon J_1 \to \widetilde{X}$ in such a way that $\widetilde{a}_1(s_0) = \widetilde{x}$ and $p \circ \widetilde{a}_1 = a_1$. After this, we obtain $\widetilde{a}_2\colon J_2 \to \widetilde{X}$ such that $p \circ \widetilde{a}_2 = a_2$ and $\widetilde{a}_2(s_*) = \widetilde{a}_1(s_*)$, which is possible because $p(\widetilde{a}_1(s_*)) = a_1(s_*) = a_2(s_*)$. Then we define $\widetilde{a}\colon J \to \widetilde{X}$ by $\widetilde{a}|J_1 = \widetilde{a}_1$ and $\widetilde{a}|J_2 = \widetilde{a}_2$. The existence of $\widetilde{a}$ in the general case reduces to the two particular cases considered because, by the continuity of $a\colon J \to X$ and the compactness of J, there exists a decomposition $J = J_1 \cup \ldots \cup J_n$ of J as the union of consecutive intervals, in such a way that $a(J_i) \subset V_i$, a distinguished neighborhood, for $i = 1, 2, \ldots, n$. The uniqueness results from Proposition 6.3. □

Now we prove that if a local homeomorphism has the unique path lifting property then the lifting $\widetilde{a}$ depends continuously of a and the initial point $\widetilde{x} = \widetilde{a}(0)$. For this purpose, we present an appropriate description of the compact-open topology for paths.

Let X be a topological space and $C(I; X)$ be the set of paths $a\colon I \to X$. Given the open sets $U_1, \ldots, U_n$ in X and a partition $0 = t_0 < t_1 < \ldots < t_n = 1$, we use the notation

$$A(t_0, t_1, \ldots, t_n; U_1, \ldots, U_n)$$

to represent the set of all paths $a\colon I \to X$ such that $a([t_{i-1}, t_i]) \subset U_i$ for $i = 1, \ldots, n$. These sets constitute the basis for a topology. From now on, the symbol $C(I;X)$ means the topological space obtained by taking this topology in the set of paths $a\colon I \to X$. We remark (but we will not use this fact) that if the topology of X comes from a metric d then the topology that we have just defined in $C(I;X)$ is induced by the metric $d(a,b) = \sup_{0\leq s\leq 1} d(a(s), b(s))$.

The following remark can be easily verified: If B is a basis of open sets in X then the open sets $A(t_0, t_1, \ldots, t_n; U_1, \ldots, U_n)$, comprised only of open sets U_i belonging to the basis B, also constitute a basis for $C(I;X)$. This fact is used in the proof of the proposition below.

Proposition 6.8. *Let $f\colon X \to Y$ be a local homeomorphism with the unique path lifting property. Given a path $a\colon I \to Y$ and a point $x \in X$ with $f(x) = a(0)$, there exists a unique path $\widetilde{a}\colon I \to X$ such that $\widetilde{a}(0) = x$ and $f \circ \widetilde{a} = a$. The lifted path $\widetilde{a}$ depends continuously on a and the initial point x. More precisely: let $\Omega \subset C(I;Y) \times X$ be the subspace whose elements are the pairs (a,x) such that $a(0) = f(x)$. Then the map $L\colon \Omega \to C(I;X)$, given by $L(a,x) = \widetilde{a}$, is continuous.*

Proof. Consider in X the basis B whose elements are the open sets U which are homeomorphically mapped by f onto open sets $V \subset Y$. Let $A = A(t_0, t_1, \ldots, t_n; U_1, \ldots, U_n)$ be an open set of the corresponding basis in $C(I;X)$, containing the path $\widetilde{a}$, and set $V_i = f(U_i)$ and $\varphi_i = (f|U_i)^{-1}$. Then the set $A(t_0, t_1, \ldots, t_n; V_1, \ldots, V_n)$ is a neighborhood of the path $a = f \circ \widetilde{a}$. We state that if the path $b\colon I \to Y$ belongs to this neighborhood and if $x' \in U_1$ then $\widetilde{b} = L(b, x')$ belongs to A. In fact, for $i = 1, 2, \ldots, n$, we have $\widetilde{b}([t_{i-1}, t_i]) = \varphi_i b([t_{i-1}, t_i]) \subset U_i$, by the uniqueness of the lifting of the restriction $b|[t_{i-1}, t_i]$ from the initial point $\widetilde{b}(t_{i-1})$. This concludes the proof. □

We will obtain the homotopy lifting property as a consequence of Proposition 6.8. In Chapter 1, we interpreted a homotopy $H\colon Z \times I \to Y$ as a path in the space $C(Z;Y)$. Now we use a dual interpretation. To each homotopy $H\colon Z \times I \to Y$ we associate a map $h\colon Z \to C(I;Y)$ which associates to each point $z \in Z$ the path $h_z\colon I \to Y$, defined by $h_z(t) = H(z,t)$. Imagining $Z \times I$ as a cylinder, union of vertical line segments, the path h_z is the restriction of H to the vertical segment $z \times I$.

Proposition 6.9. *$H\colon Z \times I \to Y$ is continuous if, and only if, $h\colon Z \to C(I,Y)$ is continuous.*

Proof. Suppose that h is continuous. Given $(z_0, t_0) \in Z \times I$, let V be a neighborhood of $H(z_0, t_0)$ in Y. We must obtain a neighborhood U of

z_0 in Z and in interval $J \subset I$, containing t_0 as an interior point (in I), such that $H(U \times J) \subset V$. By the continuity of the path h_{z_0}, there exists a closed interval $J \subset I$, containing t_0 as an interior point (in I), such that $H(z_0, t) = h_{z_0}(t) \in V$ for all $t \in J$. Let A the set of all paths $a \in C(I;Y)$ such that $a(J) \subset V$. It is obvious that A is a neighborhood of h_{z_0} in $C(I;Y)$. By the continuity of h, there exists a neighborhood U of z_0 in Z such that $h_z \in A$ for every $z \in U$; that is, $H(z,t) \in V$ for all $z \in U$ and $t \in J$.

Conversely, let H be continuous. To prove the continuity of h, let $z_0 \in Z$ and consider the basic neighborhood

$$A(t_0, \ldots, t_n; V_1, \ldots, V_n)$$

of h_{z_0}. We must find a neighborhood U of z_0 in such a way that $z \in U$ and $t_{i-1} \leq t \leq t_i$ imply $H(z,t) \in V_i (1 \leq i \leq n)$. Now, $H^{-1}(V_i)$ is a neighborhood of $z_0 \times [t_{i-1}, t_i]$ in $Z \times I$. Since $[t_{i-1}, t_i]$ is compact, there exists an open set $U_i \subset Z$, containing z_0, such that $U_i \times [t_{i-1}, t_i] \subset H^{-1}(V_i)$. We set $U = U_1 \cap \ldots \cap U_n$. This concludes the proof. □

Proposition 6.10. *Let $\varphi \colon X \to Y$ be a local homeomorphism with the unique path lifting property. Given a homotopy $H \colon Z \times I \to Y$ between two continuous maps $f, g \colon Z \to Y$, if f has a lifting $\widetilde{f} \colon Z \to X$, then g also admits a lifting $\widetilde{g} \colon Z \to X$, which is homotopic to $\widetilde{f}$. More precisely, the homotopy H admits a unique lifting $\widetilde{H} \colon Z \times I \to X$ such that $\widetilde{H}(z, 0) = \widetilde{f}(z)$ for every $z \in Z$; $\widetilde{g}$ is then defined by $\widetilde{g}(z) = \widetilde{H}(z, 1)$.*

Proof. Let $h \colon Z \to C(I;Y)$ be obtained from H as in the previous proposition. For each $z \in Z$, let $\widetilde{h}_z = L(h_z, \widetilde{f}(z))$ be the unique path lifting h_z with origin at the point $\widetilde{f}(z)$. Since L is continuous, we see that $z \mapsto \widetilde{h}_z$ defines a continuous map $\widetilde{h} \colon Z \to C(I;X)$ and therefore a homotopy $\widetilde{H} \colon Z \times I \to X$, satisfying $\widetilde{H}(z, 0) = \widetilde{h}_z(0) = \widetilde{f}(z)$ and $\varphi \widetilde{H}(z,t) = \varphi \widetilde{h}_z(t) = h_z(t) = H(z,t)$. This concludes the proof. □

Proposition 6.11. *Let $f \colon X \to Y$ be a local homeomorphism with the unique path lifting property. If the paths $a, b \colon I \to Y$, with the same endpoints y_0, y_1, are homotopic then their liftings $\widetilde{a}, \widetilde{b} \colon I \to X$, starting at the same point x_0, end at the same point x_1 and, moreover, they are homotopic.*

Proof. Let $H \colon I \times I \to Y$ be a homotopy between a and b and $\widetilde{H} \colon I \times I \to X$ be the lifting of H such that $\widetilde{H}(s, 0) = \widetilde{a}(s)$ for every $s \in I$. Since $f(\widetilde{H}(0,t)) = H(0,t) = y_0$ and $f(\widetilde{H}(1,t)) = H(1,t) = y_1$ do not depend on t, it follows that $\widetilde{H}(0,t) = x_0$ and $\widetilde{H}(1,t) = x_1$ also do not depend on t

because the fibers $f^{-1}(y_0)$ and $f^{-1}(y_1)$ are discrete. The path $s \mapsto \widetilde{H}(s,1)$ in X is a lifting of b starting at x_0. It follows from the unique path lifting property that $\widetilde{H}(s,1) = \widetilde{b}(s)$ for all $s \in I$. Thus, we have $\widetilde{H}\colon \widetilde{a} \cong \widetilde{b}$. □

Corollary 6.8. *Let $f\colon X \to Y$ be a local homeomorphism with unique path lifting property. If the closed path $a\colon I \to Y$ is homotopic to a constant, any lifting $\widetilde{a}\colon I \to X$ is closed and homotopic to a constant.*

Note that in Proposition 6.11, nothing prevents the paths a and b from being closed. Also note that the fact that $\widetilde{a}$ is closed does not imply that a is homotopic to a constant. (See Proposition 7.2.)

Example 6.12. Now we use Proposition 6.11 to exhibit an example of a space whose fundamental group is not abelian. Our space X is the union of two circles with a point x_0 in common. It is convenient to think of X as the union of the great parallel of the torus and a meridian of the same torus, which cuts the parallel at the point x_0. We denote by a a closed path that covers the parallel homeomorphically with the exception, of course, of the endpoints that are mapped onto x_0; b denotes an analogous path defined over the meridian. We introduce a covering space $\widetilde{X}$, which is the subset of the plane sketched in Figure 6.6. To obtain $\widetilde{X}$ we take on the rectangular axis, starting from the origin, four segments of length 1. From the free endpoint of each of the four segments, we take three segments of length 1/2, parallel to the axis. From the free endpoint of each of these twelve segments, we take three segments of length 1/4, and so on. The covering space $\widetilde{X}$ is the union of the segments (an infinite number) thus constructed. The covering map $p\colon \widetilde{X} \to X$ sends each horizontal segment onto a and each vertical segment onto b in such a way that the increasing order of

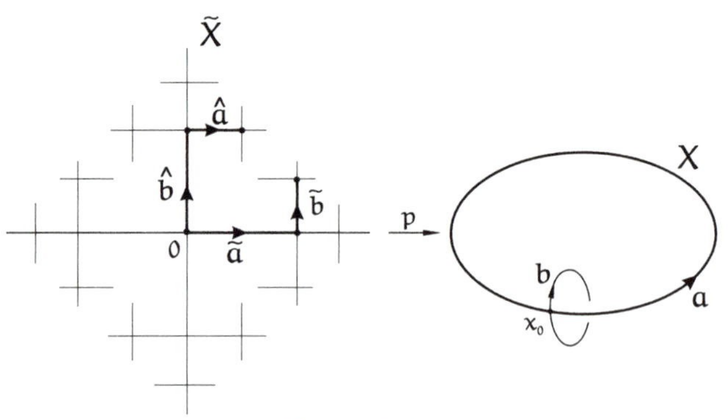

Figure 6.6.

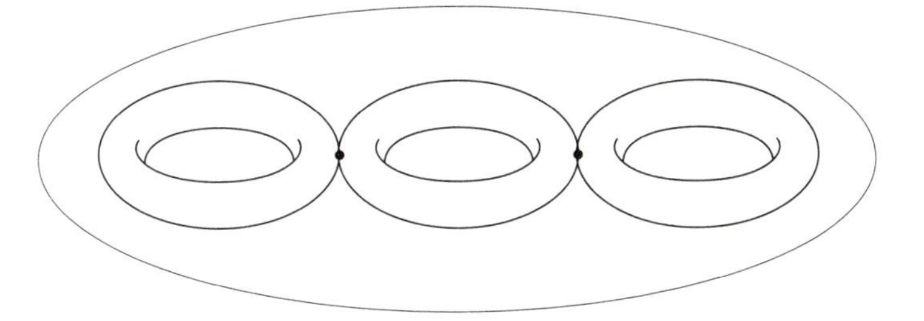

Figure 6.7.

the coordinate that varies in each one of these segments agrees with the orientations of the paths a and b respectively and that the endpoints are mapped onto x_0. ◁

Now we show that the closed paths ab and ba, with bases at the point x_0, are not homotopic in X. For this, we just have to consider their respective liftings $\widetilde{a}\widetilde{b}$ and $\widehat{b}\widehat{a}$ in $\widetilde{X}$, with origin at the point O. The final point of $\widetilde{a}\widetilde{b}$ is $(1, 1/2)$, while $\hat{b}\hat{a}$ ends at the point $(1/2, 1)$. If ab and ba were homotopic in X, their liftings from the point O would end at the same point of $\widetilde{X}$, because of Proposition 6.11.

The space X above is known as the *figure 8* space because it is homeomorphic to the graphical sign of the digit eight. Now we can exhibit other spaces with non-abelian fundamental group. For example, the union of a list (finite or infinite) of circles, each one of them with a point in common only with the previous and the following circles in the list. If the number of circles is ≥ 2, such an space has the figure eight as a retract, hence its fundamental group is not abelian. (See Proposition 2.10.) Also, a compact non-orientable surface of genus $g \geq 2$ has a non abelian fundamental group because it admits as a retract a union of g circles with $g - 1$ points of tangency. (In Figure 6.7, $g = 3$.)

We should also mention the complement of a set of two points in $\mathbb{R}^2$. This space has the same homotopy type of the figure eight space, so its fundamental group is nonabelian.

Example 6.13. The fundamental group of the figure eight space is generated by the homotopy classes $\alpha = [a]$ and $\beta = [b]$. This follows from Proposition 2.11. In fact, by denoting the figure eight space by X, we have $X = X_1 \cup X_2$, where X_1 and X_2 are circles with a point x_0 in common. We cannot directly apply the mentioned proposition because neither X_1 nor X_2 are open sets in X. But, if we take the points $x_1 \in X_1$ and $x_2 \in X_2$, both different from x_0, and set $U = X - \{x_1\}$, $V = X - \{x_2\}$, we see that the inclusions $X_2 \to U$ and $X_1 \to V$ are homotopy equivalences. From this, it follows that the homotopy class of a is a generator of the infinite cyclic group $\pi_1(V)$ and the class of b generates the infinite cyclic group $\pi_1(U)$. The same proposition, applied to $X = U \cup V$, states that $\pi_1(X, x_0)$ is generated by α and β. In fact, we may state a sharper result: The generators α and β are *free*; that is, no monomial of the type $\alpha^m \beta^n \alpha^p \ldots$, product of a finite number of alternating powers of α and β, can be reduced to the neutral element of $\pi_1(X, x_0)$ except when the exponents $m, n, p, \ldots \in \mathbb{Z}$ are all null. This fact will be proved in the next chapter. $\triangleleft$

Example 6.14. It follows from Example 6.12 that a compact orientable surface of genus > 1 does not admit a topological group structure. In fact, from Proposition 2.12, the fundamental group of a topological group is abelian. It remains to consider the compact orientable surfaces of genus 0 and 1. The torus $T = S^1 \times S^1 = \mathbb{R}^2/\mathbb{Z}^2$ has genus 1, and it is obviously a topological group; the sphere S^2 has genus 0, and it does not admit a structure of topological group, but for a completely different reason, which can be explained as follows: Suppose that S^2 is a topological group, with neutral element e. By fixing a point $a \in S^2$, close to e but satisfying $a \neq e$, we would have $a \cdot x \neq -x$ and $a \cdot x \neq x$ for all $x \in S^2$. Then, by defining $v\colon S^2 \to \mathbb{R}^3$ by setting $v(x) = \langle x, a \cdot x \rangle x - a \cdot x$, the map would be continuous, with $v(x) \neq 0$ and $\langle x, v(x) \rangle = 0$ for all $x \in S^2$, in contradiction with Proposition 4.4. Thus, we conclude that the torus is the only compact orientable surface that admits a topological group structure. In Example 7.18, we show that no compact surface (orientable or not), except the torus, can be a topological group. $\triangleleft$

Proposition 6.12. *Let $f\colon X \to Y$ a local homeomorphism with the unique path lifting property. If X is pathwise connected and Y is simply connected, then f is a homeomorphism.*

Proof. We just have to prove that f (which is already continuous, open, and surjective) is also injective. Consider $x_0, x_1 \in X$ such that $f(x_0) = f(x_1)$. Now take a path $\widetilde{a}\colon I \to X$ whose initial point is x_0 and final point is x_1. The path $a = f \circ \widetilde{a}$ is closed in Y, therefore it is homotopic to a constant. By Corollary 6.8, its lifting $\widetilde{a}$ is closed; hence, $x_0 = x_1$. $\square$

Corollary 6.9. *Let $f\colon X \to Y$ be a local homeomorphism with the unique path lifting property. If Y is simply connected and locally pathwise connected, then f maps each connected component of X homeomorphically onto Y.*

Since the space X is locally homeomorphic to Y, it is locally pathwise connected. Thus, every connected component $C \subset X$ is pathwise connected and open. Hence, $p|C$ is a local homeomorphism and, as can be easily proved, $p|C\colon C \to Y$ has the unique path lifting property. It follows from Proposition 6.12 that $p|C$ is a homeomorphism from C onto Y.

It follows from Proposition 6.12 that every pathwise connected covering of a simply connected space is a homeomorphism.

For example, let $U \subset \mathbb{R}^n$ be an open connected and bounded set. Given a class C^1 map $f\colon U \to \mathbb{R}^n$, suppose that $f'(x)\colon \mathbb{R}^n \to \mathbb{R}^n$ is, for all $x \in U$, an isomorphism. By the Inverse Function theorem, f is a local homeomorphism. It may happen that f is not a covering of the open set $V = f(U)$. But if f is such that $x_k \to x \in \partial U \Rightarrow f(x_k) \to y \in \partial V$, then f extends to a continuous map $\bar{f}\colon \overline{U} \to \overline{V}$ such that $\bar{f}(\partial U) \subset \partial V$. The map f is, in this case, proper, and therefore, it is a covering $f\colon U \to V$. If we know that V is simply connected (for example, V convex), then we may conclude that f is injective and therefore, it is a C^1 diffeomorphism from U onto V.

Analogously, let $f\colon M^m \to N^m$ be a class C^1 map where M^m and N^m are differentiable surfaces (without boundary) of dimension m. Suppose that the derivative $f'(x)\colon T_xM \to T_{f(x)}N$ is an isomorphism at each point $x \in M$. If M is compact and connected and N is simply connected, then f is bijective and therefore, it is a diffeomorphism from M onto N. The case where M is not compact will be covered in Section 5.

Example 6.15. A local homeomorphism from a connected space onto a simply connected space may not be injective (if it is not a covering map). For example: Let $X = \mathbb{C} - \{1, -1\}$, $Y = \mathbb{C}$ and define $f\colon X \to Y$ by $f(z) = z^3 - 3z$. Since $f'(z) \neq 0$ for all $z \in X$, we see that f is a local homeomorphism (Inverse Function theorem), that it is surjective because the values 2 and -2, of the polynomial $z^3 - 3z$ at the points 1 and -1, are also attained at the points 2 and -2, which belong to X. But f is not injective, even though its image $Y = \mathbb{C}(= \mathbb{R}^2)$ is simply connected. In fact, $f(0) = f(\sqrt{3}) = f(-\sqrt{3}) = 0$. Therefore, $f\colon X \to \mathbb{C}$ is a non-injective local homeomorphism onto a simply connected space. ◁

Corollary 6.10. *Let $f\colon X \to Y$ be a local homeomorphism with the unique path lifting property and $V \subset Y$ an open connected and pathwise locally connected set, such that every closed path contained in V is homotopic to a constant in Y. Then each connected component of $U = f^{-1}(V)$ is mapped homeomorphically by f onto V.*

If V were simply connected, we would have a particular case of Corollary 6.9. In the general case, we just have to observe that, with the assumed hypothesis on V, the lifting of every closed path contained in V is a closed path. This is enough to assure that $f|U$ is injective in each connected component of $U = f^{-1}(V)$ and the proof of Corollary 6.9 applies, word by word.

Corollary 6.10 says that in every set V to the above type we can define several "branches" of the inverse of f, one branch for each connected component of $f^{-1}(V)$. For example, if we take $f\colon \mathbb{R}^2 \to \mathbb{R}^2 - \{0\}$ given by $f(z) = e^z$, we reobtain the well known fact that in each simply connected region $V \subset \mathbb{R}^2 - \{0\}$ it is possible to define an infinite number of branches of the logarithm.

Corollary 6.11. *Let $p\colon \widetilde{X} \to X$ be a covering map. If an open set $V \subset X$ is connected and locally pathwise connected and, moreover, every closed path in V is homotopic to a constant in X, then V is a distinguished neighborhood.*

A topological space X is called *semi-locally simply connected* when every point $x \in X$ has a neighborhood V such that every closed path in V is homotopic to a constant in X.

Important cases of semi-locally simply connected spaces are the topological manifolds and the polyhedra. In fact, in these spaces, every point has a simply connected neighborhood, so these spaces are actually locally simply connected.

Example 6.16. We give now an example of a space Y that is semi-locally simply connected, but contains a point that does not have any simply connected neighborhood.

We start with a space X, pathwise connected, which is not semi-locally simply connected: For each $n \in \mathbb{N}$, let X_n be the circle of center $(0, 1/n)$ and radius $1/n$, in the plane. (X_n is tangent to the x-axis at the origin.)

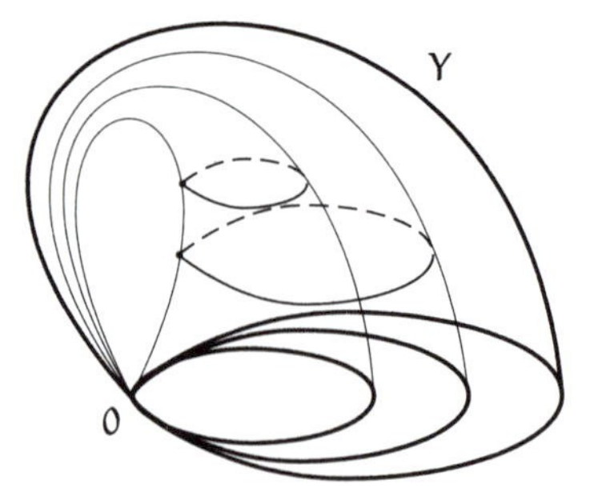

Figure 6.8.

We take $X = \bigcup_n X_n$. The space Y is obtained by taking the cone with base X and identifying the vertex of this cone with the origin O, the tangency point of the circles X_n (see Figure 6.8). ◁

Proposition 6.13. *Let X be a locally pathwise connected and semi-locally simply connected space. A map $p\colon \widetilde{X} \to X$ is a covering if, and only if, it is a local homeomorphism with the unique path lifting property.*

Proof. The "If" part is the Corollary 6.10 above. The "Only if" part is Proposition 6.7. □

Corollary 6.12. *Let X be a locally pathwise connected and semi-locally simply connected space. If $p\colon \widetilde{X} \to X$ and $q\colon \widehat{X} \to \widetilde{X}$ are covering maps, the composite map $p \circ q\colon \widehat{X} \to X$ is also a covering map.*

In fact, it is obvious that if the maps p and q have the unique path lifting property, then the composite map $p \circ q$ also has the property.

Remark. If p has a finite number of leaves then, as we can see from the definition, its composite map $p \circ q$, with another covering map q, is still a covering map, even without imposing to one of the spaces (and therefore to all of them) the condition of being semi-locally simply connected.

6.4.1 An Application

Let G, H be topological groups. A *local homomorphism* from G to H is a *continuous* map $f\colon U \to H$, defined in a neighborhood U of the neutral element $e \in G$, such that if $x, y, x \cdot y \in U$, then $f(x \cdot y) = f(x) \cdot f(y)$. As an application of Proposition 6.12, we prove the following.

If the group G is simply connected and locally pathwise connected, then every local homomorphism $f\colon U \to H$, from G into a topological group H, extends to a continuous homomorphism $\bar{f}\colon G \to H$.

In fact, restricting f, if necessary, we may suppose that its domain U is pathwise connected. Let $A \subset G \times H$ be the subgroup of the product $G \times H$ generated by the graph of f. We define on A the topological group topology according to which a fundamental system of neighborhoods of the neutral element is given by the sets $\widetilde{V} = \{(x, f(x)); x \in V\}$, where $V \subset U$ is a neighborhood of the neutral element. Let $p\colon A \to G$ be the restriction of the projection $\pi_G\colon G \times H \to G$. The continuous homomorphism p maps the graph $\widetilde{U}$ of f homeomorphically onto U. Since G is connected, and therefore generated by U, p is surjective (and a local homeomorphism). By Corollary 6.10, $p\colon A \to G$ is a covering. Now, G is simply connected and A is connected, because it is generated by the connected neighborhood $\widetilde{U}$ = graph of f. Hence, p is a homeomorphism from A onto G. The inverse homeomorphism $p^{-1}\colon G \to A$ is given by $p^{-1}(x) = (x, \bar{f}(x))$. The continuous homomorphism $\bar{f}\colon G \to H$, thus defined, is the extension of f that we have been searching for.

6.5 Differentiable Coverings

First, we examine what happens when a surjective local homeomorphism $f\colon X \to Y$ does not have the path lifting property.

This means that there exist $x \in X$ and a path $a\colon I \to Y$ such that $a(0) = f(x)$ but a cannot be lifted to a path in X starting at the point x. We suppose that X is Hausdorff, which gives us the uniqueness of the liftings that might there exist.

Since f is a local homeomorphism, for $\varepsilon > 0$ sufficiently small the restriction $a|[0,\varepsilon]$ has a lifting starting at the point x. Therefore, there exists a number r, $0 < r \leq 1$, such that, for all r' with $0 < r' < r$, the path $a|[0,r']$ has a lifting starting at the point x but $a|[0,r]$ does not have. This means (because of the uniqueness of the lifting) that $a|[0,r)$ has a lifting $\widetilde{a}\colon [0,r) \to X$ but, when $s \to r$, $\widetilde{a}(s)$ does not have an adherence value (hence no limit) in X. (In fact, if $x' \in X$ were an adherence value of $\widetilde{a}(s)$ when $s \to r$, the continuity of f would imply that $f(x')$ would be the adherence value for $a(s)$ when $s \to r$ and therefore $f(x') = a(r)$. Then, by taking a neighborhood of x' mapped homeomorphically by f onto a neighborhood of $a(r)$, we would conclude that $a|[0,r]$ would have a lifting.) The non-existence of an adherence value of $\widetilde{a}(s)$ in X when $s \to r$, results, in particular, that the set $\{\widetilde{a}(s); 0 \leq s < r\}$ is closed in X, while its image by f, that is, $\{a(s); 0 \leq s < r\}$, is not closed in Y. Hence we can state the

Proposition 6.14. *Let X be a Hausdorff space. If a surjective local homeomorphism $f\colon X \to Y$ is a closed map then f has the unique path lifting property. In particular if, moreover, Y is locally pathwise connected and semi-locally simply connected, then f is a covering map.*

Remark. Under very general conditions, if a local homeomorphism $f\colon X \to Y$ is a closed map, then f is proper; that is, $f^{-1}(y)$ is a finite set, for all $y \in Y$. (See Proposition A.6 in the Appendix.)

We provide now a sufficient condition in order that a map be a covering within the scope of the differential calculus.

Proposition 6.15. *Let $f\colon \mathbb{R}^m \to \mathbb{R}^m$ be a map of class C^1, whose values are contained in a open connected set $Y \subset \mathbb{R}^m$. Suppose that there exists*

a covering of Y *by open sets* V*, and to each of these sets is associated a number* $\varepsilon_V > 0$*, in such a way that* $f(x) \in V$ *implies* $|f'(x) \cdot u| \geq \varepsilon_V \cdot |u|$ *for all* $u \in \mathbb{R}^m$*. Then* $f(\mathbb{R}^m) = Y$ *and* $f\colon \mathbb{R}^m \to Y$ *is a covering map.*

Proof. First we show that if $a\colon [0,1] \to Y$ is a path of class C^1 in Y and $b\colon [0,1) \to \mathbb{R}^m$ is such that $f(b(s)) = a(s)$, $0 \leq s < 1$, then b is of class C^1 and there exists $\lim_{s\to 1} b(s)$ in $\mathbb{R}^m$. The fact that $b \in C^1$ follows easily from the fact that f is a local diffeomorphism of class C^1. Next, let $y_1 = a(1)$ and consider $V \ni y_1$, $\varepsilon_V > 0$ as in the statement of the proposition. There exists $\delta > 0$ such that $1 - \delta < s < 1 \Rightarrow f(b(s)) = a(s) \in V$ and therefore $|f'(b(s)) \cdot b'(s)| \geq \varepsilon_V \cdot |b'(s)|$. On the other hand, $f'(b(s)) \cdot b'(s) = a'(s)$, hence $|b'(s)| \leq |a'(s)|/\varepsilon_V$ when $1 - \delta < s < 1$. Since the interval $[0,1]$ is compact and a if of class C^1, there exists $A > 0$ such that $|a'(s)| \leq A \cdot \varepsilon_V$ for all $s \in [0,1]$. Therefore, if $1 - \delta < s_1$, $s_2 < 1$, we have:

$$|b(s_2) - b(s_1)| = \left| \int_{s_1}^{s_2} b'(s)ds \right| \leq |s_2 - s_1| \cdot A.$$

By the Cauchy criterion in the complete metric space $\mathbb{R}^m$, it follows that the limit $\lim_{s\to 1} b(s)$ exists.

Now we prove that every rectilinear path contained in Y, starting at an arbitrary point $y_0 \in f(\mathbb{R}^m)$, can be lifted from any point $x_0 \in f^{-1}(y_0)$. In fact, it this were not true, there would exist a rectilinear path $a(s) = (1-s)y_0 + sy_1$ in Y such that the restriction $a|[0,1)$ would have a lifting $b\colon [0,1) \to \mathbb{R}^m$, with $b(0) = x_0$, and such that the limit $\lim_{s\to 1} a(s)$ would not exist. But this contradicts what we have proved above.

Now we verify that $f(\mathbb{R}^m)$ is a closed subset of the open set Y. In fact, every y_1 that belongs to the closure of $f(\mathbb{R}^m)$, relatively to Y, can be connected to a point $y_0 \in f(\mathbb{R}^m)$ by a rectilinear path contained in Y, which can be lifted to $\mathbb{R}^m$, in such a way that $y_1 \in f(\mathbb{R}^m)$. Since Y is connected and $f(\mathbb{R}^m)$ is obviously open, it follows that $f(\mathbb{R}^m) = Y$.

Therefore, every rectilinear path in Y can be lifted, and the proposition follows from Lemma 6.1 below. □

Lemma 6.1. *Let* $Y \subset \mathbb{R}^m$ *be an open set. In order to verify the path lifting property relative to a local homeomorphism* $f\colon X \to Y$*, it suffices to consider the rectilinear paths in* Y*; that is, the paths* $a\colon I \to Y$ *defined by* $a(s) = (1-s)y_0 + sy_1$*.*

Proof. Suppose initially that Y is convex. If every rectilinear path a in Y has a lifting $\widetilde{a}$ in X, starting at an arbitrary point x in $f^{-1}(a(0))$ then, naturally, $\widetilde{a}$ is unique and depends continuously on a. (See Propositions 6.2

and 6.8.) Therefore, if $a: I \to Y$ is any path in Y, let $a_t: I \to Y$, $0 \le t \le 1$, the rectilinear path that connects $a(0)$ to $a(t)$; that is, $a_t(s) = (1-s)a(0) + sa(t)$, $0 \le s \le 1$. Given $x \in f^{-1}(a(0))$, let $\widetilde{a}_t$ be the lifting of a_t that starts at the point x. We define a path $\widetilde{a}: I \to X$ by setting $\widetilde{a}(t) = \widetilde{a}_t(1)$. Then $\widetilde{a}$ is a lifting of a starting at the point x.

In the general case, Y can be covered by open balls and the above argument shows that every path contained in one of these balls can be lifted to X. Now we observe that any path in Y can be decomposed in a finite sequence of smaller paths, such that each one of them is contained in an open ball and therefore, it can be lifted. It follows that the whole path can be lifted, which proves the lemma. □

Corollary 6.13. *Let $f: \mathbb{R}^m \to \mathbb{R}^m$ a map of class C^1. If there exists $\alpha > 0$ such that $|f'(x) \cdot v| \ge \alpha \cdot |v|$ for all x and every v in $\mathbb{R}^m$, then f is a bijection and therefore, it is a diffeomorphism from $\mathbb{R}^m$ onto itself.*

In fact, take $Y = V = \mathbb{R}^m$ and $\varepsilon_V = \alpha$ in the proposition. Then f is a covering of $\mathbb{R}^m$. Since $\mathbb{R}^m$ is simply connected, it follows from Proposition 6.12 that f is a bijection and therefore it is a diffeomorphism.

Corollary 6.14. *Let $f: \mathbb{R}^m \to \mathbb{R}^m$ be a map of class C^1 such that $|f'(x) \cdot v| = |v|$ for all x and every v in $\mathbb{R}^m$. Then f is an isometry; that is, $|f(x) - f(y)| = |x - y|$ for any $x, y \in \mathbb{R}^m$. (As we know from linear algebra, this implies that there exist a linear orthogonal transformation $T: \mathbb{R}^m \to \mathbb{R}^m$ and a vector $c \in \mathbb{R}^m$ such that $f(x) = T \cdot x + c$ for every $x \in \mathbb{R}^m$.)*

In fact, by Corollary 6.13, f is a diffeomorphism. The Mean Value theorem applied to f gives us $|f(x) - f(y)| \le |x - y|$ for any $x, y \in \mathbb{R}^m$. The same theorem applied to f^{-1} gives us $|x - y| \le |f(x) - f(y)|$. Hence, f is an isometry.

Proposition 6.15 can be stated in a global scope, by considering Riemannian manifolds instead of open sets in Euclidean space. The proof follows precisely the same argument, substituting the rectilinear paths by geodesics and the convex subsets of $\mathbb{R}^m$ by geodesically convex sets. Corollary 6.13 is valid only for complete, simply connected manifolds (same proof) and Corollary 6.14 is false (see $\xi: \mathbb{R} \to S^1$, $\xi(t) = e^{it}$). The statement of the global version of Proposition 6.15 follows:

Proposition 6.16. *Let M^m, N^m be Riemannian manifolds of the same dimension m, with M^m complete and N^m connected. Suppose that there exists a map $f: M \to N$, of class C^1, and a covering of N by open sets*

V, and to each of these open sets it is associated a number $\varepsilon_V > 0$ such that $x \in M$, $f(x) \in V \Rightarrow |f'(x) \cdot u| \geq \varepsilon_V \cdot |u|$ for every $u \in T_xM$. Then $f\colon M \to N$ is a covering map.

This proposition, with the same proof, is still valid for Banach manifolds, by omitting the sentence "of the same dimension m" and by requiring that the derivative $f'(x)\colon T_xM \to T_yN$, $y = f(x)$, be an isomorphism, for every $x \in M$.

6.6 Exercises

1. Give the following examples:

   a) A continuous bijection which is not a local homeomorphism;

   b) A continuous surjective map $f\colon \mathbb{R} \to \mathbb{R}$ such that $f^{-1}(y)$ is discrete for every $y \in \mathbb{R}$ but f is not locally injective;

   c) A counter-example to Proposition 6.2 with Z disconnected.

2. If $p\colon \widetilde{X} \to X$ and $q\colon \widetilde{Y} \to Y$ are coverings, then the map $p \times q\colon \widetilde{X} \times \widetilde{Y} \to X \times Y$, defined by $(p \times q)(x, y) = (p(x), q(y))$, is also a covering.

3. Consider the covering map $p\colon \widetilde{X} \to X$, and let $Y \subset X$ be an arbitrary subset. Set $\widetilde{Y} = p^{-1}(Y)$ and $q = p|\widetilde{Y}$. Show that $q\colon \widetilde{Y} \to Y$ is a covering.

4. Let $p\colon \widetilde{X} \to X$ be a covering where the base X is connected and locally connected. For every connected component $C \subset \widetilde{X}$, we have $p(C) = X$. Conclude that $p|C\colon C \to X$ is a covering.

5. Contrary to the function $f\colon S^1 \to S^1$, $f(z) = z^2$, there does not exist a continuous map $\varphi\colon \mathbb{R} \to \mathbb{R}$ such that $\varphi^{-1}(y)$ has exactly two points, for every $y \in \mathbb{R}$.

6. Given the polynomial $p(z) = 2z^3 - 9z^2 + 12z + 1$, obtain two finite subsets $F_1 \subset \mathbb{C}$ e $F_2 \subset \mathbb{C}$ such that $p\colon \mathbb{C} - F_1 \to \mathbb{C} - F_2$ is a covering with three leaves.

7. Let X be a connected space and G a properly discontinuous group of homeomorphisms of X. Suppose that a continuous map $f\colon X \to X$ has the following property: For every $x \in X$, there exists $g \in G$ such that $f(x) = gx$. Prove that f is a homeomorphism.

8. Let E be an equivalence relation in the Hausdorff space X, such that the quotient map $\varphi\colon X \to X/E$ is open. Prove that X/E is a Hausdorff

space if, and only if, the graph $T = \{(x, y) \in X \times X; xEy\}$ is a closed set in $X \times X$.

9. Let G be any homeomorphism group of the space X. Consider in G the discrete topology and show that the map $\varphi\colon G \times X \to X$, $\varphi(g, x) = gx$, is a covering.

10. Let $f\colon X \to Y$ be a local homeomorphism with the unique path lifting property. Suppose that Y is simply connected and locally pathwise connected. Show that, for every $x_0 \in X$ with $f(x_0) = y_0$, there exists a section $\sigma\colon Y \to X$, such that $\sigma(y_0) = x_0$. Derive from this again Proposition 6.12.

11. Consider the neighborhood $U = \{e^{it}; -\pi < t < \pi\}$ of the neutral element of S^1, and define the local homomorphism $f\colon U \to S^1$, $f(e^{it}) = e^{it/2}$. Show that f does not extend to a continuous homomorphism $\bar{f}\colon S^1 \to S^1$.

12. Let $p\colon X \to X$ be a covering and a, $b\colon I \to X$ be freely homotopic closed paths. If b, has a closed lifting $\widetilde{b}$, then a also has a closed lifting $\widetilde{a}$, which is freely homotopic to $\widetilde{b}$.

13. Let G be a simply connected and locally pathwise connected topological group. If a connected topological group K is locally isomorphic to G, then K is isomorphic to a quotient of G by a discrete subgroup H (necessarily contained in the center of G).

14. A compact and connected hypersurface $M^n \subset \mathbb{R}^{n+1}$ of class C^∞ whose Gaussian curvature is different from zero at every point is diffeomorphic to the sphere S^n. (The Gaussian curvature is the Jacobian determinant of the normal map $M^n \to S^n$.)

15. Given the covering $p\colon \widetilde{X} \to X$ and the continuous map $f\colon Z \to X$, let $\widetilde{Z} = \{(z, \widetilde{x}) \in Z \times \widetilde{X}; f(z) = p(\widetilde{x})\}$. Prove that the map $q\colon \widetilde{Z} \to Z$, defined by $q(z, \widetilde{x}) = z$ is a covering. Prove also that f admits a continuous lifting $\sigma\colon Z \to \widetilde{X}$ if, and only if, there exists a continuous section $\sigma\colon Z \to \widetilde{Z}$ for q.

16. Show that Exercise 3 follows from Exercise 15.

17. Let U be the set of quaternions $w = t + xi + yj + zk$ where $t > 0$ and X the set of real quaternions ≤ 0. By setting $V = \mathbb{R}^4 - X$, prove that the map $f\colon U \to V$, defined by $f(w) = w^2$, is a surjective proper local diffeomorphism, and conclude that f is a diffeomorphism (global) from U onto V.

18. Let H be a locally pathwise connected, closed subgroup of the connected group G. If G/H is simply connected, prove that H is connected.

(Suggestion: Consider H_0, a connected component of the neutral element. Observe that H_0 is the normal subgroup of H, H/H_0 is a discrete subgroup of G/H_0, and the natural projection from G/H_0 onto its quotient by H/H_0 induces a covering $G/H_0 \to G/H$; hence, $H = H_0$.)

19. Let $p\colon \widetilde{X} \to X$ be a covering with $\widetilde{X}$ connected and $p^{-1}(x)$ finite, for every $x \in X$. If there exists a continuous map $f\colon \widetilde{X} \to \mathbb{R}$, injective in each fiber $p^{-1}(x)$, then p is a homeomorphism.

Chapter 7

Covering Maps and Fundamental Groups

7.1 The Conjugate Class of a Covering Map

Given a covering map $p\colon \widetilde{X} \to X$, take $\widetilde{x} \in \widetilde{X}$ and set $x = p(\widetilde{x})$. We use the notation $H(\widetilde{x})$ to represent the image of the homomorphism $p_{\#}\colon \pi_1(\widetilde{X}, \widetilde{x}) \to \pi_1(X, x)$, induced by the covering projection p.

The subgroup $H(\widetilde{x}) \subset \pi_1(X, x)$ is, as we show in this chapter, the most important algebraic tool to characterize the covering $p\colon \widetilde{X} \to X$.

If $\widetilde{X}$ is simply connected, then $H(\widetilde{x}) = \{0\}$ for all $\widetilde{x} \in \widetilde{X}$. The converse is also true and follows from Proposition 1 below.

First we should recall that, by fixing an element g in a group G, the map $x \mapsto g \cdot x \cdot g^{-1}$ is an automorphism of G, called *conjugation* by g. If H is a subgroup of G, its image by this automorphism is the subgroup $g \cdot H \cdot g^{-1} = \{g \cdot x \cdot g^{-1}; x \in H\}$, isomorphic to H, called a *conjugate subgroup of* H. The conjugate class of H in G is the set of all subgroups $g \cdot H \cdot g^{-1}$, conjugate of H, obtained when we vary g in G. The subgroup H is said to be *normal* when $g \cdot H \cdot g^{-1} = H$ for every $g \in G$; that is, when its conjugate class has only one element, namely, the group H itself. This happens, for example, when G is abelian.

Proposition 7.1. *Let $p\colon \widetilde{X} \to X$ be a covering. For any $x_0 \in X$, $\widetilde{x}_0 \in p^{-1}(x_0)$, the induced homomorphism $p_{\#}\colon \pi_1(\widetilde{X}, \widetilde{x}_0) \to \pi_1(X, x_0)$ is injective. If $\widetilde{X}$ is pathwise connected, then, when $\widetilde{x}$ varies in the fiber $p^{-1}(x_0)$, the image $H(\widetilde{x}) = p_{\#}\pi_1(\widetilde{X}, \widetilde{x})$ describes all conjugate classes of the subgroup $H(\widetilde{x}_0)$.*

Proof. By Corollary 6.8, the induced homomorphism $p_{\#}\colon \pi_1(\widetilde{X}, \widetilde{x}_0) \to \pi_1(X, x_0)$ is injective. Given any other point $\widetilde{x} \in p^{-1}(x_0)$, there exists in $\widetilde{X}$ a path $\widetilde{c}$, with origin $\widetilde{x}$ and final point $\widetilde{x}_0$. Then $c = p \circ \widetilde{c}$ is a closed path in X, with base at the point x_0. By Proposition 2.5, every element $\alpha \in \pi_1(\widetilde{X}, \widetilde{x})$ has the form $\alpha = [\widetilde{c}\widetilde{b}\widetilde{c}^{-1}]$, where $[\widetilde{b}] \in \pi_1(\widetilde{X}, \widetilde{x}_0)$. Hence, $p_{\#}(\alpha) = \gamma p_{\#}([\widetilde{b}])\gamma^{-1}$, where $\gamma = [c]$. From this, $H(\widetilde{x}) = \gamma \cdot H(\widetilde{x}_0) \cdot \gamma^{-1}$.

Conversely, let $H = \gamma \cdot H(\widetilde{x}_0) \cdot \gamma^{-1}$ be any conjugate subgroup of $H(\widetilde{x}_0)$ in $\pi_1(X, x_0)$, and set $\gamma = [c]$. By lifting the closed path c^{-1} from the point $\widetilde{x}_0$, we obtain a path $\widetilde{c}^{-1}$ in $\widetilde{X}$, whose final point we denote by $\widetilde{x}$. Then $\widetilde{x} \in p^{-1}(x_0)$ and the path $\widetilde{c}$, in $\widetilde{X}$, begins at $\widetilde{x}$ and ends at $\widetilde{x}_0$, with $p \circ \widetilde{c} = c$. From what we just saw, this gives us $H(\widetilde{x}) = \gamma \cdot H(\widetilde{x}_0) \cdot \gamma^{-1}$ and therefore $H = H(\widetilde{x})$. □

Proposition 7.2. *Consider a covering $p\colon \widetilde{X} \to X$. Let $a, b\colon I \to X$ be paths that start at the same point x and end at the same point y, and $\widetilde{a}, \widetilde{b}\colon I \to \widetilde{X}$ their liftings from a point $\widetilde{x} \in \widetilde{X}$. In order that $\widetilde{a}(1) = \widetilde{b}(1)$, it is necessary and sufficient that $[ab^{-1}] \in H(\widetilde{x})$.*

Proof. Assume that $[ab^{-1}] \in H(\widetilde{x})$. Then the lifting $\widetilde{c}$ of the path ab^{-1} from the point $\widetilde{x}$ is closed. The paths $\widetilde{a}, \widetilde{b}\colon I \to \widetilde{X}$, defined by $\widetilde{a}(s) = \widetilde{c}(s/2)$ and $\widetilde{b}(s) = \widetilde{c}(1 - s/2)$, start at the point $\widetilde{x}$, end at the same point $\widetilde{c}(1/2)$ and are, respectively, liftings of a and b. The converse is obvious. □

The following corollary will be important later on.

Corollary 7.1. *Let $p\colon \widetilde{X} \to X$ be a covering. Given a closed path $a\colon I \to X$, with base at the point x, its lifting $\widetilde{a}\colon I \to \widetilde{X}$, from a point $\widetilde{x} \in p^{-1}(x)$, is closed if, and only if, $[a] \in H(\widetilde{x})$.*

This follows from Proposition 7.2 by taking $b = e_x$.

Corollary 7.2. *Let $p\colon \widetilde{X} \to X$ be a covering, with $\widetilde{X}$ simply connected. A closed path $a\colon I \to X$ is homotopic to a constant if, and only if, some of its lifting $\widetilde{a}\colon I \to \widetilde{X}$ is closed. (And then all of its liftings are closed.)*

More generally: *Still supposing that $\widetilde{X}$ is simply connected, consider the paths a, $b\colon I \to X$ with the same endpoints x_0, x_1. Let $\widetilde{a}$, $\widetilde{b}\colon I \to \widetilde{X}$ be their liftings from the point $\widetilde{x}_0 \in p^{-1}(x_0)$. We have $a \cong b$ if, and only if, $\widetilde{a}$ and $\widetilde{b}$ end at the same point $\widetilde{X}$.*

Corollary 7.3. *Let $p\colon \widetilde{X} \to X$ be a covering, with $\widetilde{X}$ pathwise connected. By fixing a point $x_0 \in X$, the following statements are equivalent:*

1. *For some* $\widetilde{x}_0 \in p^{-1}(x_0)$, *the subgroup* $H(\widetilde{x}_0) \subset \pi_1(X, x_0)$ *is normal;*

2. *The subgroups* $H(\widetilde{x}) \subset \pi_1(X, x_0)$, *when* $\widetilde{x}$ *varies in* $p^{-1}(x_0)$, *are normal and they are all the same;*

3. *Given a closed path* $a: I \to X$, *with base at* x_0, *either all of the liftings of* a *from the points* $\widetilde{x} \in p^{-1}(x_0)$ *are closed or none of them is closed.*

In fact, by Proposition 7.2, Condition 3 above is equivalent to stating that $[a] \in H(\widetilde{x}_a) \Leftrightarrow [a] \in H(\widetilde{x})$ for all $\widetilde{x} \in p^{-1}(x_0)$. This means that all of the groups $H(\widetilde{x})$ are equal, when $\widetilde{x}$ varies in the fiber $p^{-1}(x_0)$. But these groups constitute a conjugate class; hence, they are equal if, and only if, one of them is normal, and therefore all of them are normal.

When $\widetilde{X}$ is pathwise connected and one of the conditions in the above corollary is satisfied (and therefore, all of the conditions are satisfied), we say that $p: \widetilde{X} \to X$ is a *regular covering.*

Remark. When $\widetilde{X}$ and X are pathwise connected, the reader can easily prove that the regularity of the covering $p: \widetilde{X} \to X$ does not depend on the point $x_0 \in X$ fixed above.

When the covering $p: \widetilde{X} \to X$ is regular, we may use $H(x_0)$, instead of $H(\widetilde{x}_0)$, in order to identify the image $p_{\#}\pi_1(\widetilde{X}, \widetilde{x}_0)$, $\widetilde{x}_0 \in p^{-1}(x_0)$.

Example 7.1. If the fundamental group of X is abelian, then every covering $p: \widetilde{X} \to X$, with $\widetilde{X}$ pathwise connected, is regular. ◁

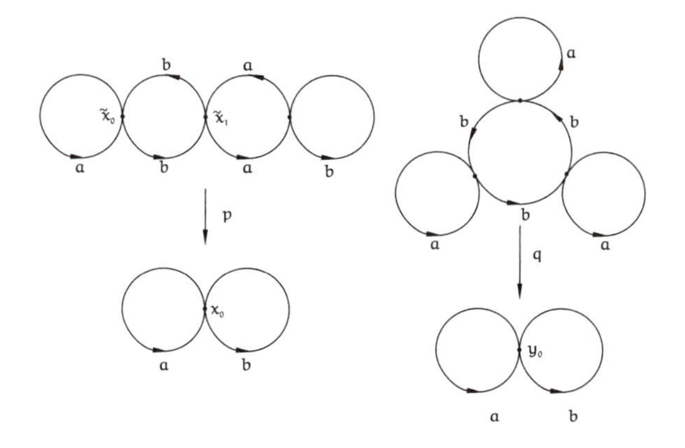

Figure 7.1.

Example 7.2. Let's consider now the two coverings of the figure eight space shown in Figure 7.1. Each one of them is a covering with three leaves. (We will see in what follows that every covering with two leaves is regular; therefore, this example is the simplest possible.) ◁

In each of these examples, the space $\widetilde{X}$ is a union of circles. In the example on the left, the circles on the extremities are applied homeomorphically onto the two circles that form the figure eight space, in the manner indicated by the letters and arrows, while each one of the circles in the middle covers twice the corresponding circle of the base. In the example on the right, the central circle of the space $\widetilde{Y}$ covers three times the circle b of the base, while each one of the three outside circles is mapped homeomorphically onto the other circle of the base. The covering $p\colon \widetilde{X} \to X$ on the left is not regular. In fact, the lifting of the closed path a is open if it starts at the point $\widetilde{x}_1$ and it is closed if its initial point is $\widetilde{x}_0$. On the other hand, the covering $q\colon \widetilde{Y} \to Y$ on the right is regular. In fact, the liftings of a are always closed and the liftings of b are always open. Since $\pi_1(Y, y_0)$ is generated by $[a]$ and $[b]$, Condition 3 of Corollary 7.3 holds.

Example 7.3. Let G be a properly discontinuous group of homeomorphisms of the pathwise connected topological space X. (See Section 6.3.) The quotient map $\pi\colon X \to X/G$ is a regular covering. In fact, choose $x_0, x_1 \in X$ such that $\pi(x_0) = \pi(x_1)$. Then $x_1 = gx_0$, $g \in G$. By the uniqueness of the lifting, the paths $\widetilde{a}$, $\hat{a}\colon I \to X$, with $\widetilde{a}(0) = x_0$, $\hat{a}(0) = x_1$, are liftings of the same path in X/G if, and only if, $\hat{a}(s) = g(\widetilde{a}(s))$ for every $s \in I$. Hence, $\hat{a}$ is closed if, and only if, $\widetilde{a}$ is closed. ◁

Before stating the next proposition, we recall some concepts from algebra.

Let S be an arbitrary set and G a group. A *right action* of the group G on the set S is a map $S \times G \to S$, which maps each pair $(x, g) \in S \times G$ to an element $xg \in S$, in such a way that the following conditions hold:

1. $x(gh) = (xg)h$;

2. $x \cdot e = x$, for any $x \in S$; $e =$ neutral element of G.

In this case, we say that the group G *operates* on *acts* on the right in the set S. A left action is defined in a similar way.

If G acts on the right in the set S, the *orbit* of an element $x \in S$ is the set $xG = \{xg; g \in G\}$. The group G is said to *operate transitively* in S when the orbit of an element of S (and therefore of all elements of S) is the set S itself. This means that, given any two elements $x, y \in S$, there exists $g \in G$ such that $y = xg$.

When G acts on the right in S, given an element $x \in S$, the set $H(x) = \{g \in G; xg = x\}$ is a subgroup of G, called *isotropy group* (or *stabilizer*) of the point x. If $y = xh$ then $yg = y \Leftrightarrow x(hgh^{-1}) = x$; that is, $H(x) = h \cdot H(y) \cdot h^{-1}$. In sum: If two elements $x, y \in S$ belong to the same orbit of G then their isotropy groups are conjugate.

Suppose that G acts transitively on the right in the set S. By fixing a point $x_0 \in S$, the map $\varphi\colon G \to S$, given by $\varphi(g) = x_0 g$ is surjective and satisfies $\varphi(g) = \varphi(h) \Leftrightarrow hg^{-1} \in H(x_0)$, where $H(x_0)$ is the isotropy group of x_0. Therefore, by passing to the quotient, φ induces a bijection

$$\overline{\varphi}\colon G/H(x_0) \to S.$$

In particular, the cardinal number of S is equal to the index $[G\colon H(x_0)]$ of the subgroup $H(x_0)$ in G; that is, the cardinal number of the set $G/H(x_0)$ of the cosets $H(x_0) \cdot g$, $g \in G$.

Proposition 7.3. *Let $p\colon \widetilde{X} \to X$ be a covering, with $\widetilde{X}$ pathwise connected. For each $x \in X$, the fundamental group $\pi_1(X, x)$ acts transitively on the right in the fiber $p^{-1}(x)$. The isotropy group of each point $\widetilde{x} \in p^{-1}(x)$ is $H(\widetilde{x}) = p_\# \pi_1(\widetilde{X}, \widetilde{x})$.*

Proof. Given $\alpha \in \pi_1(X, x)$ and $\widetilde{x} \in p^{-1}(x)$, we define $\widetilde{x}\alpha \in p^{-1}(x)$ as follows: we choose $a \in \alpha$, lift the path a from the initial point $\widetilde{x}$, take the final point $\widetilde{y}$ of this lifting and set $\widetilde{x}\alpha = \widetilde{y}$. It is easy to verify that this procedure defines (without ambiguities) an operation of $\pi_1(X, x)$ on the right in the fiber $p^{-1}(x)$. We have $\widetilde{x}\alpha = \widetilde{x}$ if, and only if, the lifting of the path a, from $\widetilde{x}$, is closed. By Corollary 7.1, this occurs if, and only if, $\alpha \in H(\widetilde{x})$. The transitivity results from the fact that $\widetilde{X}$ is pathwise connected: Given $\widetilde{x}, \widetilde{y} \in p^{-1}(x)$, let $\widetilde{a}$ be a path in $\widetilde{X}$ starting at $\widetilde{x}$ and ending at $\widetilde{y}$. Then $a = p \circ \widetilde{a}$ is a closed path in X with base at the point x. Let $\alpha = [a]$. It is obvious that $\widetilde{y} = \widetilde{x}\alpha$. □

Corollary 7.4. *If $\widetilde{X}$ is pathwise connected then, for any $\widetilde{x} \in \widetilde{X}$, and $x = p(\widetilde{x})$, the number of leaves of p is equal to the index of the subgroup $H(\widetilde{x}) \subset \pi_1(X, x)$.*

Corollary 7.5. *If $\widetilde{X}$ is pathwise connected, every covering $p\colon \widetilde{X} \to X$ with two leaves is regular.*

In fact, every subgroup of index two is normal.

Corollary 7.6. *Let $\widetilde{X}$ be pathwise connected. The covering projection $p\colon \widetilde{X} \to X$ is a homeomorphism if, and only if, the induced homeomorphism $p_\#$ is an isomorphism.*

In fact, this is the condition we should have in order that the number of leaves be one.

Corollary 7.7. *If $\widetilde{X}$ is simply connected, then the number of leaves of the covering is equal to the number of elements of $\pi_1(X,x)$. When these two numbers are finite, the equality between them implies that $\widetilde{X}$ is simply connected.*

Remark. The permutations of the fiber $p^{-1}(x)$ of the form $\widetilde{x} \mapsto \widetilde{x}\cdot\alpha$, where $\alpha \in \pi_1(X,x)$, form a group $M(x)$, called the *monodromy group* of the covering $p\colon \widetilde{X} \to X$ at the point x. For all $x \in X$, $M(x)$ is a homomorphic image of $\pi_1(X,x)$. More precisely, we have

$$M(x) \approx \pi_1(X,x)/H_0,$$

where

$$H_0 = \bigcap_{p(\widetilde{x})=x} H(\widetilde{x}).$$

If the covering is regular, we have $H_0 = H(\widetilde{x})$ for all $\widetilde{x} \in p^{-1}(x)$.

7.2 The Fundamental Lifting Theorem

In this section, we show how the fundamental group allows us to give an algebraic answer to the topological problem of knowing whether a continuous map $f\colon Z \to X$, taking values at the base of a covering, admits a lifting $\widetilde{f}\colon Z \to \widetilde{X}$. It is convenient here to use pairs (X, x_0); that is, spaces with a base point.

Proposition 7.4. *Let $p\colon \widetilde{X} \to X$ be a covering, of the pathwise connected space X. Let Z be a connected and locally pathwise connected space (hence pathwise connected) and $f\colon (Z,z_0) \to (X,x_0)$ a continuous map. Given $\widetilde{x}_0 \in p^{-1}(x_0)$, in order that f have a lifting $\widetilde{f}\colon (Z,z_0) \to (\widetilde{X},\widetilde{x}_0)$, it is necessary and sufficient that $f_\#\pi_1(Z,z_0) \subset H(\widetilde{x}_0)$.*

Proof. If there exists $\widetilde{f}\colon (Z,z_0) \to (\widetilde{X},x_0)$ continuous such that $p \circ \widetilde{f} = f$, then, considering the homomorphisms induced by f, p and $\widetilde{f}$, we see that the diagram below is commutative:

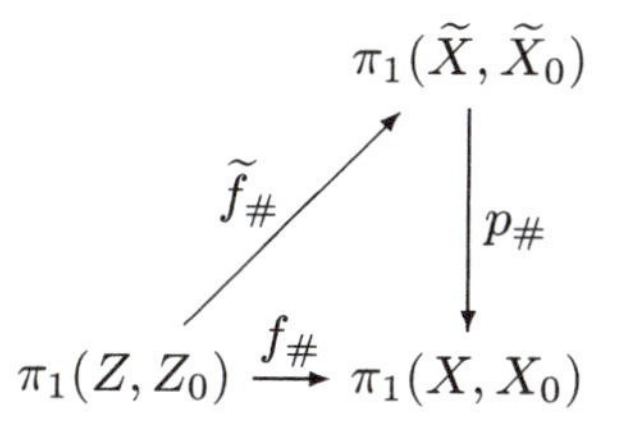

From $f_\# = p_\# \circ \widetilde{f}_\#$, it follows that the image of $f_\#$ is contained in the image of $p_\#$, which is $H(\widetilde{x}_0)$. Hence the inclusion is necessary for the existence of $\widetilde{f}$.

Conversely, suppose that $f_\#\pi_1(Z, z_0) \subset H(\widetilde{x}_0)$. We define $\widetilde{f}\colon Z \to \widetilde{X}$, by setting $\widetilde{f}(z_0) = \widetilde{x}_0$ and, for an arbitrary $z \in Z$, we take a path $a\colon I \to Z$, from z_0 to z, we denote by $\widetilde{a}\colon I \to \widetilde{X}$ the lifting of $f \circ a\colon I \to X$ from the point $\widetilde{x}_0$ and we set $\widetilde{f}(z) = \widetilde{a}(1)$. Now let us show that $\widetilde{f}$ is well defined. In fact, if $b\colon I \to Z$ is another path from z_0 to z, then ba^{-1} is a closed path with base z_0. From this, $(f \circ b)(f \circ a)^{-1} = f \circ (ab^{-1})$ is a closed path, with base x_0, whose homotopy class belongs to the image of $f_\#$ and therefore (because of the hypothesis) to $H(\widetilde{x}_0)$. From this it results that the paths $\widetilde{a}$ e $\widetilde{b}$, liftings of $f \circ a$ and $f \circ b$, respectively, from $\widetilde{x}_0$, end at the same point. (See Proposition 7.2.) Evidently, we have $p \circ \widetilde{f} = f$. It remains to prove only that $\widetilde{f}$ is continuous at an arbitrary point $z \in Z$. Here we use the fact that Z is locally pathwise connected. Let V be a neighborhood of $\widetilde{f}(z)$ in $\widetilde{X}$. We may suppose that $p|V$ is a homeomorphism onto a neighborhood U of $f(z)$ in X. Let W be a pathwise connected neighborhood of z in Z such that $f(W) \subset U$. We claim that $\widetilde{f}(W) \subset V$. This will prove the continuity of $\widetilde{f}$ at the point z. We know that $\widetilde{f}(z)$ is the final point of a path $\widetilde{a}$ in $\widetilde{X}$ that starts at $\widetilde{x}_0$, with $p \circ \widetilde{a} = f \circ a$, where a is a path in Z, starting at z_0 and ending at z. Given $w \in W$, we take a path b in W, starting at z and ending at w. Since $p|V$ is a homeomorphism onto U, there exists a path $\widetilde{b}$ in V, starting at $\widetilde{f}(z)$ and ending at a certain point $v \in V$, with $p \circ \widetilde{b} = f \circ b$. Then $\widetilde{a}\widetilde{b}$ is a path in $\widetilde{X}$, that starts at $\widetilde{x}_0$, such that $p \circ (\widetilde{a}\widetilde{b}) = (p \circ \widetilde{a})(p \circ \widetilde{b}) = (f \circ a)(f \circ b) = f \circ (ab)$. Since ab connects z_0 to w in Z, it follows from the definition of $\widetilde{f}$ that $\widetilde{f}(w) = (\widetilde{a}\widetilde{b})(1) = v$; therefore, $\widetilde{f}(w) \in V$. □

Second Proof of Proposition 7.4. The experienced topologist, faced with an argument where, in order to define a map, he has to make arbitrary choices that turn out to be irrelevant, always suspects that such map might somehow be obtained by passing to the quotient. In the present case, the suspicion is true, as we show now.

Let $C(Z; z_0) \subset C(I; Z)$ be the subspace (in the compact-open topology) whose elements are the paths with origin at the point z_0. A similar interpretation is given to the notations $C(X; x_0)$ and $C(\widetilde{X}; \widetilde{x}_0)$. The following three maps are continuous: $f_*\colon C(Z; z_0) \to C(X; x_0)$, $L\colon C(X; x_0) \to C(\widetilde{X}; \widetilde{x}_0)$, and $v\colon C(\widetilde{X}; \widetilde{x}_0) \to \widetilde{X}$, given by $f_*(a) = f \circ a$, $L(c) = \widetilde{c} =$ lifting of c from $\widetilde{x}_0$, and $v(\widetilde{a}) = \widetilde{a}(1)$. Therefore the composite map $\hat{f} = v \circ L \circ f_*\colon C(Z; z_0) \to \widetilde{X}$ is also continuous. On the other hand, the surjection $u\colon C(Z; z_0) \to Z$, defined by $u(a) = a(1)$, is continuous and, moreover, it is open (which follows from the fact that Z is pathwise locally connected). Hence u is a quotient map. Now we observe that the hypothesis on the image of $f_\#$ gives us: $u(a) = u(a') \Rightarrow \hat{f}(a) = \hat{f}(a')$. Thus, $\hat{f}$ is compatible with the equivalence relation defined by u. By passing to the quotient, there exists therefore a unique continuous map $\widetilde{f}\colon Z \to \widetilde{X}$ such that $\widetilde{f} \circ u = \hat{f}$. The map $\widetilde{f}$ is the lifting of f we have been looking for. □

Corollary 7.8. *Let X be pathwise connected and Z be simply connected and locally pathwise connected. Every continuous map $f\colon (Z, z_0) \to (X, x_0)$ admits a lifting $\widetilde{f}\colon (Z, z_0) \to (\widetilde{X}, \widetilde{x}_0)$, where $\widetilde{x}_0 \in p^{-1}(x_0)$ is chosen arbitrarily.*

The above corollary explains why it is always possible to lift a path: I is simply connected.

As an application, we use Proposition 7.4 to establish the conditions under which a continuous complex function has a continuous logarithm.

Example 7.4. (Logarithm of a function) Let $U \subset \mathbb{C}$ be an open connected set and $f\colon U \to \mathbb{C} - \{0\}$ a continuous map. In order that there exists $g\colon U \to \mathbb{C}$ continuous such that $f(z) = e^{g(z)}$ for all $z \in U$, it is necessary and sufficient that, for every closed path $a\colon I \to U$, the number or turns of the closed path $f \circ a\colon I \to \mathbb{C} - \{0\}$ around the point 0 be equal to zero. When f is holomorphic, the function g is necessarily holomorphic. In fact, the condition on the number of turns of the path $f \circ a$ means that the induced homomorphism $f_\#\colon \pi_1(U, u_0) \to \pi_1(\mathbb{C} - \{0\})$ is null. Considering the covering map $p\colon \mathbb{C} \to \mathbb{C} - \{0\}$, given by $p(z) = e^z$, we see that the induced homomorphism $p_\#\colon \pi_1(\mathbb{C}) \to \pi_1(\mathbb{C} - \{0\})$ is null because $\pi_1(\mathbb{C}) = \{0\}$. Thus, f has a lifting relatively to p if, and only if, $f_\# = 0$. Now, the fact that g is a lifting of f relatively to p means that $f(z) = e^{g(z)}$ for all $z \in U$. Since g is continuous, and p, f are holomorphic, with $p'(z) \neq 0$, it follows from the Inverse Function Theorem that $f(z) = p(g(z))$

for every z implies that g is holomorphic. Note, in particular, that when U is simply connected, $f_{\#}$ is always null; hence, every continuous function (respectively holomorphic) that is non-null in a simply connected domain always admits a continuous logarithm. (It is usual to call every continuous function g such that $f(z) = e^{g(z)}$ a "branch of $\log f(z)$".) ◁

In an analogous way, we study the existence of the k-th root of a map in the example below.

Example 7.5. Considering the covering map $p\colon \mathbb{C}-\{0\} \to \mathbb{C}-\{0\}, p(z) = z^k$, $k \in \mathbb{N}$, (Example 6.6), we show that, *given a continuous function* $f\colon U \to \mathbb{C} - \{0\}$, *defined in an open, connected set* $U \subset \mathbb{C}$, *there exists* $g\colon U \to \mathbb{C} - \{0\}$ *continuous (called a "branch of* $\sqrt[k]{f(z)}$*") such that* $f(z) = g(z)^k$ *for every* $z \in U$ *if, and only if, every closed path* $a\colon I \to U$ *is mapped by* f *in a path* $f \circ a\colon I \to \mathbb{C} - \{0\}$, *whose number of turns around the origin* 0 *is a multiple of* k. (Again, if f is holomorphic, g is also holomorphic.) We just have to observe that, for the covering $p(z) = z^k$, the image of the homomorphism $p_{\#}$ is the subgroup of $\pi_1(\mathbb{C} - \{0\}) = \mathbb{Z}$ formed by the multiples of k. The condition on the number of turns of the path $f \circ a$ means that the image of the homomorphism $f_{\#}\colon \pi_1(U, u_0) \to \pi_1(\mathbb{C} - \{0\})$ is contained in the image of $p_{\#}$. Hence, such a condition is necessary and sufficient in order that f have a lifting relative to p. Now, g is such a lifting if, and only if, $f(z) = g(z)^k$ for all $k \in U$. Again, we remark that, in particular, if U is simply connected, for every continuous function $f\colon U \to \mathbb{C} - \{0\}$, there always exists a branch of $\sqrt[k]{f(z)}$ defined on U. ◁

The following proposition, which is also a direct application of Proposition 7.4, expresses, grosso modo, that every covering of a topological group is still a topological group.

Proposition 7.5. *Let* G *be a locally pathwise connected topological group, with neutral element* e. *Given a covering* $p\colon \widetilde{G} \to G$, *with* $\widetilde{G}$ *connected, and a point* $\widetilde{e} \in p^{-1}(e)$, *there exists a unique topological group structure in* $\widetilde{G}$, *such that* $\widetilde{e}$ *is the neutral element and* p *is a homomorphism.*

Proof. Let $p{\cdot}p\colon (\widetilde{G} \times \widetilde{G}, (\widetilde{e}, \widetilde{e})) \to (G, e)$ be the continuous map defined by $(p{\cdot}p)(\widetilde{x}, \widetilde{y}) = p(\widetilde{x}){\cdot}p(\widetilde{y})$. The essential point consists in proving that $p{\cdot}p$ has a lifting $m\colon (\widetilde{G} \times \widetilde{G}, (\widetilde{e}, \widetilde{e})) \to (\widetilde{G}, \widetilde{e})$. The image of the induced homomorphism $(p \cdot p)_{\#}$ is the set of homotopy classes of all paths in G of the form $a{\cdot}b$, where $[a]$ and $[b]$ belong to the image of $p_{\#}$. Thus, the lifting $m\colon \widetilde{G} \times \widetilde{G} \to \widetilde{G}$, with $m(\widetilde{e}, \widetilde{e}) = \widetilde{e}$ and $p(m(\widetilde{x}, \widetilde{y})) = p(\widetilde{x}){\cdot}p(\widetilde{y})$, exists. For the sake of simplicity, we write $m(\widetilde{x}, \widetilde{y}) = \widetilde{x}{\cdot}\widetilde{y}$. Therefore, we have a continuous multiplication in $\widetilde{G}$, which turns p into an homomorphism. It remains to verify that it turns $\widetilde{G}$ into a group, in which $\widetilde{e}$ is the neutral element. First, the continuous maps $\widetilde{x} \mapsto \widetilde{x}\cdot\widetilde{e}$ and $\widetilde{x} \mapsto \widetilde{x}$, of $\widetilde{G}$ into $\widetilde{G}$, are liftings of the same map p, which have the same value at the point $\widetilde{e}$. Hence, they coincide; that is, we have $\widetilde{x} \cdot \widetilde{e} = \widetilde{x}$ for all $\widetilde{x} \in \widetilde{G}$. In an analogous way, we see that $\widetilde{e} \cdot \widetilde{x} = \widetilde{x}$; therefore, $\widetilde{e}$ is neutral element for the multiplication of $\widetilde{G}$. The associativity is proved by observing that the maps $(\widetilde{x}, \widetilde{y}, \widetilde{z}) \mapsto (\widetilde{x}\cdot\widetilde{y})\cdot\widetilde{z}$ and $(\widetilde{x}, \widetilde{y}, \widetilde{z}) \mapsto \widetilde{x}\cdot(\widetilde{y}\cdot\widetilde{z})$ are liftings of the map $(\widetilde{x}, \widetilde{y}, \widetilde{z}) \mapsto p(\widetilde{x})\cdot p(\widetilde{y})\cdot p(\widetilde{z})$, which coincide at the point $(\widetilde{e}, \widetilde{e}, \widetilde{e})$. By virtue of Proposition 2.12, the map $\widetilde{x} \mapsto p(\widetilde{x})^{-1}$ has a lifting $i\colon \widetilde{G} \to \widetilde{G}$, such that $i(\widetilde{e}) = \widetilde{e}$ and $p(i(\widetilde{x})) = p(\widetilde{x})^{-1}$, so $p(i(\widetilde{x})\cdot\widetilde{x}) = e$, and from this $i(\widetilde{x})\cdot\widetilde{x} \in p^{-1}(e)$ for all $\widetilde{x} \in \widetilde{G}$. Since $\widetilde{G}$ is connected and the fiber $p^{-1}(e)$ is discrete, the product $i(\widetilde{x})\cdot\widetilde{x}$ is constant when $\widetilde{x}$ varies in $\widetilde{G}$. Now we observe that $i(\widetilde{e})\cdot\widetilde{e} = \widetilde{e}$. Hence, $i(\widetilde{x})\cdot\widetilde{x} = \widetilde{e}$, which gives us $i(\widetilde{x}) = \widetilde{x}^{-1}$. □

Let $p\colon \widetilde{G} \to G$ be a *homomorphic covering*; that is, a covering map that is also a homomorphism between the topological groups $\widetilde{G}$ and G.

When $\widetilde{G}$ (and therefore, G) is pathwise connected, the covering $p\colon \widetilde{G} \to G$ is regular, because $\pi_1(G)$ is abelian. Moreover, the following proposition holds.

Proposition 7.6. *Let $K = p^{-1}(e)$ be the kernel of the homomorphic covering $p\colon \widetilde{G} \to G$, where $\widetilde{G}$ is pathwise connected. There exists a natural isomorphism $\pi_1(G)/\pi_1(\widetilde{G}) \approx K$.*

Proof. Above, we are identifying $\pi_1(\widetilde{G})$ with its image $H(\widetilde{e})$ using the induced homomorphism $p_{\#}$. In order to obtain the isomorphism, we define $\varphi\colon \pi_1(G) \to K$ by setting $\varphi(\alpha) = \widetilde{a}(1)$, where $\widetilde{a}$ is the lifting, from $\widetilde{e}$, of a path $a \in \alpha$. (In the notation of Proposition 7.3, $\varphi(\alpha) = \widetilde{e}\cdot\alpha$.) We claim that φ is a homomorphism. In order to verify this, we consider operation of the group $\pi_1(G)$ as being $\alpha \cdot \beta$. (See Proposition 2.12.) If $\widetilde{a}$ and $\widetilde{b}$ are liftings, from $\widetilde{e}$, of the paths $a \in \alpha$ and $b \in \beta$ respectively, the fact that p is a homomorphism gives us $\widetilde{a\cdot b} = \widetilde{a}\cdot\widetilde{b}$. Hence $\varphi(\alpha \cdot \beta) = (\widetilde{a}\cdot\widetilde{b})(1) = \widetilde{a}(1)\cdot\widetilde{b}(1) = \varphi(\alpha)\cdot\varphi(\beta)$. The homomorphism φ is surjective, because $\pi_1(G)$ acts transitively in the fiber $p^{-1}(e) = K$. The kernel of φ is the set of elements $\alpha = [a] \in \pi_1(G)$ such that $\widetilde{a}$ is closed. By Proposition 7.2, this kernel is $\pi_1(\widetilde{G})$. By passing to the quotient, we obtain an isomorphism $\overline{\varphi}\colon \pi_1(G)/\pi_1(\widetilde{G}) \to K$. □

Corollary 7.9. *Let $p\colon \widetilde{G} \to G$ be a homomorphic covering. If $\widetilde{G}$ is simply connected, then $\pi_1(G)$ is isomorphic to the kernel $K = p^{-1}(e)$ of the homomorphism p.*

Example 7.6. The covering map $p\colon \mathbb{R} \to S^1$, given by $p(x) = e^{2\pi i x}$, is a homomorphism of the (additive) group $\mathbb{R}$ over the (multiplicative) group

S^1, with kernel $\mathbb{Z}$. Since $\mathbb{R}$ is simply connected, we reobtain the fact that $\pi_1(S^1) \approx \mathbb{Z}$. Analogously, we may consider the homomorphic covering $p\colon \mathbb{R}^n \to T^n = S^1 \times \cdots \times S^1$, $p(x_1, \ldots, x_n) = (e^{2\pi i x_1}, \ldots, e^{2\pi i x_n})$, whose kernel is $\mathbb{Z}^n$, and find that the fundamental group of the n-dimensional torus T^n is isomorphic to $\mathbb{Z}^n$. Finally, the homomorphism $\varphi\colon S^3 \to \mathrm{SO}(3)$, defined in Section 1 of Chapter 4, has kernel $\{-1, 1\} \approx \mathbb{Z}_2$ and it is surjective. Therefore, it is a homomorphic covering. Since S^3 is simply connected, it follows that $\pi_1(\mathrm{SO}(3)) \approx \mathbb{Z}_2$. $\triangleleft$

Example 7.7. We can compute again $\pi_1(\mathrm{SO}(4))$ by taking a homomorphic covering whose base is $\mathrm{SO}(4)$ and whose domain is simply connected. For this, we define $p\colon S^3 \times S^3 \to \mathrm{SO}(4)$ by associating to each pair $(x, y) \in S^3 \times S^3$ the linear transformation $p_{x,y}\colon \mathbb{R}^4 \to \mathbb{R}^4$, given by $p_{x,y}(w) = xwy^{-1}$ (quaternion multiplication). Since $|x| = |y| = 1$, the linear transformation $p_{x,y}$ preserves norms; since x and y may be connected to 1 by paths in S^3, $p_{x,y}$ can be connected to the identity transformation by a path formed by linear transformations that preserve norms. Hence $p_{x,y} \in \mathrm{SO}(4)$. It is obvious that $p\colon S^3 \times S^3 \to \mathrm{SO}(4)$, thus defined, is an infinitely differentiable homomorphism. If (x, y) belongs to the kernel of p, then $x \cdot w \cdot y^{-1} = w$ for all $w \in \mathbb{R}^4$. In particular, $x \cdot 1 \cdot y^{-1} = 1$, so $x = y$. It follows that $x \cdot w \cdot x^{-1} = w$ for all $w \in \mathbb{R}^4$. As in Section 4.1, we conclude that $x = \pm 1$. Thus, the kernel of the homomorphism $p\colon S^3 \times S^3 \to \mathrm{SO}(4)$ consists of two elements $(1, 1)$ and $(-1, -1)$. Finally, since $\dim(S^3 \times S^3) = \dim \mathrm{SO}(4) = 6$ and the kernel of p is discrete, the rank theorem, from analysis in Euclidean space, tells us that p has rank 6 and therefore, it is an open map. Since $S^3 \times S^3$ is compact and $\mathrm{SO}(4)$ is Hausdorff connected, this implies that p is surjective. The surjective homomorphism p is a covering because it has a discrete kernel. Its domain, $S^3 \times S^3$, is simply connected. Hence, $\pi_1(\mathrm{SO}(4))$ has two elements and therefore, it is isomorphic to $\mathbb{Z}_2$. $\triangleleft$

7.3 Homomorphisms of Covering Spaces

Let $p_1\colon \widetilde{X}_1 \to X$ and $p_2\colon \widetilde{X}_2 \to X$ be two coverings with the same base space X. A *homomorphism* between them is a continuous map $f\colon \widetilde{X}_1 \to \widetilde{X}_2$ such that $p_2 \circ f = p_1$. This means that the following diagram is commutative.

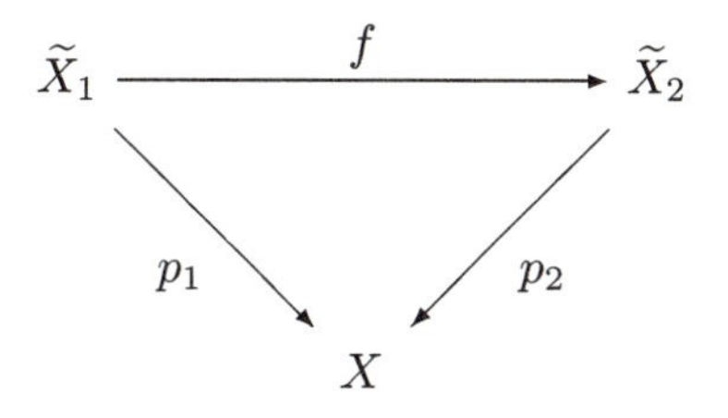

If $p_3 \colon \widetilde{X}_3 \to X$ is another covering with base space X and $g \colon \widetilde{X}_2 \to \widetilde{X}_3$ is a homomorphism, the composite map $g \circ f \colon \widetilde{X}_1 \to \widetilde{X}_3$ is still a homomorphism. We say that $f \colon \widetilde{X}_1 \to \widetilde{X}_2$ is an *isomorphism* when f is a homeomorphism such that $p_2 \circ f = p_1$. Then $f^{-1} \colon \widetilde{X}_2 \to \widetilde{X}_1$ is also an isomorphism. In this case, the coverings p_1 and p_2 are said to be *isomorphic.*

An *endomorphism* is an homomorphism of a covering into itself. Given a covering $p \colon \widetilde{X} \to X$, an endomorphism is, therefore, a continuous map $f \colon \widetilde{X} \to \widetilde{X}$ such that $p \circ f = p$.

When the endomorphism f is a homeomorphism of $\widetilde{X}$ onto itself, we say that f is an *automorphism.* The set $G(\widetilde{X}|X)$ of the covering automorphisms $p \colon \widetilde{X} \to X$ is a group under the operation of map composition. Sometimes automorphisms are called *covering transformations* or *covering translations.*

The condition $p_2 \circ f = p_1$ means that f maps each fiber $p_1^{-1}(x)$ into the fiber $p_2^{-1}(x)$. In particular, an endomorphism $f \colon \widetilde{X} \to \widetilde{X}$ maps each fiber $p^{-1}(x)$ into itself. An isomorphism f induces, for each $x \in X$, a bijection of the fiber $p_1^{-1}(x)$ onto the fiber $p_2^{-1}(x)$. An automorphism, therefore, determines a permutation of each fiber $p^{-1}(x)$.

Note that a homomorphism $f \colon \widetilde{X}_1 \to \widetilde{X}_2$ is a lifting of the continuous map $p_1 \colon \widetilde{X}_1 \to X$ with respect to the covering $p_2 \colon \widetilde{X}_2 \to X$. Thus, when $\widetilde{X}_1$ is connected, two homomorphisms that coincide at a point $\widetilde{x}_1 \in \widetilde{X}_1$ are equal.

Example 7.8. Consider the covering maps $p_1 \colon \mathbb{R}^2 \to T^2$, from the plane onto the torus and $p_2 \colon S^1 \times \mathbb{R} \to T^2$, from the cylinder onto the torus, given by $p_1(s,t) = (e^{2\pi i s}, e^{2\pi i t})$ and $p_2(z,t) = (z, e^{2\pi i t})$. The map $f \colon \mathbb{R}^2 \to S^1 \times \mathbb{R}$, from the plane onto the cylinder, given by $f(s,t) = (e^{2\pi i s}, t)$, satisfies the condition $p_2 \circ f = p_1$; hence, it is a covering homomorphism. Note that the subgroup $H_1 = \{0\} \subset \pi_1(T^2)$ is associated to the covering p_1, and the subgroup $H_2 = \mathbb{Z} \oplus \{0\}$ of $\pi_1(T^2) = \mathbb{Z} \oplus \mathbb{Z}$ is associated to the covering p_2. We have $H_1 \subset H_2$. It is this inclusion that makes it possible the existence of f. $\triangleleft$

Example 7.9. Let G be a properly discontinuous group of homeomorphisms of the connected topological space X. We claim that the group of covering automorphisms of the covering $p \colon X \to X/G$ is precisely the group G. In fact, if $g \in G$ then, for all $x \in X$, we have $p(gx) = G \cdot gx = G \cdot x = p(x)$, hence $p \circ g = p$ and from this $g \in G(X|X/G)$. Conversely, given an automorphism $f \colon X \to X$, we fix $x_0 \in X$ and take $x_1 = f(x_0)$. Then x_1 belongs to the same fiber that x_0, hence there exists $g \in G$ with $gx_0 = x_1$. Therefore f and g are liftings of p that coincide at the point x_0. Since X is connected, we have $f = g$, so $f \in G$. Thus, for example, the automorphisms of the covering $p\colon \mathbb{R}^2 \to T^2$, of the torus by the plane, $p(s,t) = (e^{2\pi i s}, e^{2\pi i t})$, are translations $(s,t) \mapsto (s+m, t+n)$ where $m, n \in \mathbb{Z}$. Another example, that generalizes this one of the torus, is the following: let $p\colon \widetilde{G} \to G$ a homomorphic covering of the connected topological group $\widetilde{G}$ onto G. The kernel $K = p^{-1}(e)$, being a normal and discrete subgroup of the connected group $\widetilde{G}$, is central, that is, their elements commute with every other element in G. The automorphisms of the covering $p\colon \widetilde{G} \to G$ are translations $f_k\colon \widetilde{G} \to \widetilde{G}$, $f_k(x) = k \cdot x = x \cdot k$, $k \in K$. In fact, the set of the translations f_k, $k \in K$, is a properly discontinuous group of homeomorphisms of $\widetilde{G}$, isomorphic to K, and the quotient space $\widetilde{G}/K$ is homeomorphic to G. ◁

Proposition 7.7. *Let $p_1\colon \widetilde{X}_1 \to X$ and $p_2\colon \widetilde{X}_2 \to X$ be coverings with the same base space X. If $\widetilde{X}_2$ is connected and locally pathwise connected, every homomorphism $f\colon \widetilde{X}_1 \to \widetilde{X}_2$ is a covering. In particular, f is surjective.*

Proof. Take $\widetilde{x}_1 \in \widetilde{X}_1$. Let $\widetilde{x}_2 = f(\widetilde{x}_1)$. If $a\colon I \to \widetilde{X}_2$ is any path starting at $\widetilde{x}_2$, we set $a_0 = p_2 \circ a$ and consider $\widetilde{a}\colon I \to \widetilde{X}_1$, the lifting of a_0 with respect to the covering p_1, starting at the point $\widetilde{x}_1$. Then $f \circ \widetilde{a}\colon I \to \widetilde{X}_2$ is a lifting of a_0 with respect to p_2, starting at the point $\widetilde{x}_2$. If follows that $f \circ \widetilde{a} = a$. In particular, $f(\widetilde{a}(1)) = a(1)$. Since $\widetilde{X}_2$ is pathwise connected, any one of its points is of the form $a(1)$, for some path a starting at $\widetilde{x}_2$. Hence f is surjective. The same argument also shows that f has the unique path lifting property. Since the relation $p_2 \circ f = p_1$ implies that the continuous map f is a local homeomorphism, the proposition is already proved in the case where one of the spaces $\widetilde{X}_1$, $\widetilde{X}_2$, X is semi-locally simply connected (and the same happens with the other two).

In the general case, let $\widetilde{x}_2 \in \widetilde{X}_2$ be an arbitrary point. Take a connected neighborhood U of the point $x_0 = p_2(\widetilde{x}_2)$, which is distinguished with respect to the coverings p_2 and p_1. (Observe that the spaces $\widetilde{X}_1$, $\widetilde{X}_2$, and X are locally homeomorphic; hence, they are locally connected.) Let V be the connected component of $p_2^{-1}(U)$ that contains the point $\widetilde{x}_2$. We claim that V is a distinguished neighborhood of $\widetilde{x}_2$, relative to f. We have

$$p_1^{-1}(U) = \bigcup_{\lambda} \widetilde{U}_\lambda,$$

a union of disjoint open sets where, for each λ, $p_1|\widetilde{U}_\lambda$ is a homeomorphism onto U. If, for some λ, we have $f(\widetilde{U}_\lambda) \cap V \neq \varnothing$, then, since the connected set $f(\widetilde{U}_\lambda)$ is contained in the set $p_2^{-1}(U)$, of which V is a connected component, it follows that $f(\widetilde{U}_\lambda) \subset V$ and from this, $f|\widetilde{U}_\lambda = (p_2|V)^{-1} \circ (p_1|\widetilde{U}_\lambda)$; hence,

f is a homeomorphism from $\widetilde{U}_\lambda$ onto V. Therefore, by setting

$$L_0 = \{\lambda; f(\widetilde{U}_\lambda) \cap V \neq \varnothing\},$$

we see that $L_0 \neq \varnothing$ (because f is surjective), that

$$f^{-1}(V) = \bigcup_{\lambda \in L_0} \widetilde{U}_\lambda$$

and that $f|\widetilde{U}_\lambda$ is a homeomorphism onto V, for all $\lambda \in L_0$, thus concluding the proof. □

Corollary 7.10. *Let $p_1 \colon \widetilde{X}_1 \to X$ and $p_2 \colon \widetilde{X}_2 \to X$ be coverings over the same base space X, with $\widetilde{X}_1$ and $\widetilde{X}_2$ connected and locally pathwise connected. A homomorphism $f \colon \widetilde{X}_1 \to \widetilde{X}_2$ is an isomorphism if, and only if, $f_\# \colon \pi_1(\widetilde{X}_1, \widetilde{x}_1) \to \pi_1(\widetilde{X}_2, f(\widetilde{x}_1))$ is surjective (and therefore an isomorphism between the fundamental groups).*

In fact, this results from Corollary 7.6, by taking into account Proposition 7.7.

In the following proposition, we have the coverings $p_1 \colon \widetilde{X}_1 \to X, p_2 \colon \widetilde{X}_2 \to X$. Given the points $\widetilde{x}_1 \in \widetilde{X}_1$ and $\widetilde{x}_2 \in \widetilde{X}_2$ with $p_1(\widetilde{x}_1) = p_2(\widetilde{x}_2) = x_0$, we denote by $H_1(\widetilde{x}_1)$ and $H_2(\widetilde{x}_2)$, respectively, the subgroups of $\pi_1(X, x_0)$ which are images of the induced homomorphisms $(p_1)_\# \colon \pi_1(\widetilde{X}_1, \widetilde{x}_1) \to \pi_1(X, x_0)$ and $(p_2)_\# \colon \pi_1(\widetilde{X}_2, \widetilde{x}_2) \to \pi_1(X, x_0)$.

Proposition 7.8. *Let $\widetilde{X}_1$, $\widetilde{X}_2$ be connected and locally pathwise connected. In order that there exists a homomorphism $f \colon \widetilde{X}_1 \to \widetilde{X}_2$ with $f(\widetilde{x}_1) = \widetilde{x}_2$ it is necessary and sufficient that $H_1(\widetilde{x}_1) \subset H_2(\widetilde{x}_2)$.*

Proof. This follows from Proposition 7.4, because a homomorphism f is a lifting of p_1 relatively to the covering p_2. □

Example 7.10. Consider the coverings $p_1 \colon S^1 \to S^1$ and $p_2 \colon S^1 \to S^1$, given by $p_1(z) = z^{12}$ and $p_2(z) = z^3$. The corresponding subgroups are $H_1 = 12\mathbb{Z}$ and $H_2 = 3\mathbb{Z}$. (Since the fundamental groups are abelian, we write H instead of $H(\widetilde{x})$.) Evidently, we have $H_1 \subset H_2$. Hence, by taking any z_1, $z_2 \in S^1$ with $p_1(z_1) = p_2(z_2)$, there exists a homomorphism $f \colon S^1 \to S^1$ such that $f(z_1) = z_2$. We choose $z_1 = z_2 = 1$. Then $f(z) = z^4$ is the homomorphism between the given coverings. This is illustrated by the commutative diagram below.

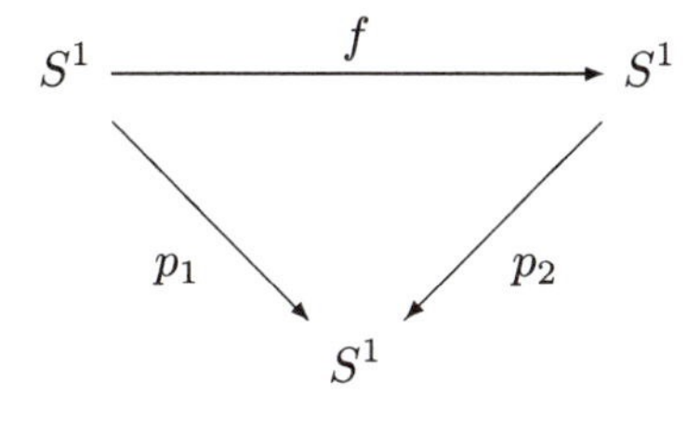

◁

Corollary 7.11. *Let $p\colon \widetilde{X} \to X$ be a covering whose domain $\widetilde{X}$ is simply connected and locally pathwise connected. For every covering $q\colon \widetilde{Y} \to X$ with $\widetilde{Y}$ connected, there exists a covering $f\colon \widetilde{X} \to \widetilde{Y}$ such that $q \circ f = p$. This is illustrated by the commutative diagram below.*

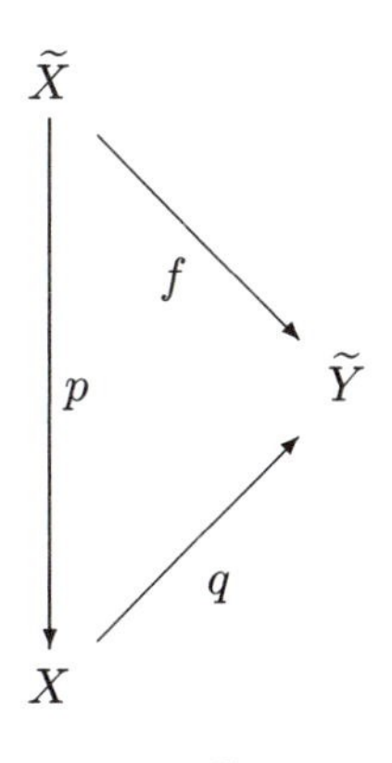

In fact, for any $\widetilde{x} \in \widetilde{X}$ and $\widetilde{y} \in \widetilde{Y}$ with $p(\widetilde{x}) = q(\widetilde{y})$ we have $\{0\} = p_{\#}\pi_1(\widetilde{X}, \widetilde{x}) \subset q_{\#}\pi_1(\widetilde{Y}, \widetilde{y})$.

Because of the above corollary, a covering $p\colon \widetilde{X} \to X$ with $\widetilde{X}$ simply connected and locally pathwise connected is called a *universal covering*, since $\widetilde{X}$ covers any other covering $\widetilde{Y}$ of the space X.

Corollary 7.12. *Under the hyphotesis of Proposition 7.8, the homomorphism $f\colon \widetilde{X}_1 \to \widetilde{X}_2$, with $f(\widetilde{x}_1) = \widetilde{x}_2$, is an isomorphism if, and only if, $H_1(\widetilde{x}_1) = H_2(\widetilde{x}_2)$.*

The "Only if" part is obvious. We just have to prove the "if" part.

In fact, in this case, Proposition 7.8 guarantees the existence of a homomorphism $\widetilde{g}\colon \widetilde{X}_2 \to \widetilde{X}_1$ with $g(\widetilde{x}_2) = \widetilde{x}_1$. Then $g \circ f\colon \widetilde{X}_1 \to \widetilde{X}_1$ is an endomorphism such that $\widetilde{x}_1$ is a fixed point, so it coincides with the identity map of $\widetilde{X}_1$. In a similar way we prove that $f \circ g\colon \widetilde{X}_2 \to \widetilde{X}_2$ is the identity, hence f is a homeomorphism.

Corollary 7.13. *Two simply connected coverings of a locally pathwise connected path are isomorphic.*

The above propositions can be summarized as follows:

A. Let X be a connected and locally pathwise connected space and $x_0 \in X$. To each covering $p\colon \widetilde{X} \to X$, with connected domain $\widetilde{X}$, there corresponds a conjugate class $\mathcal{H}(x_0)$, of subgroups of $\pi_1(X, x_0)$, that consists of the subgroups $H(\widetilde{x}) = p_{\#}\pi_1(\widetilde{X}, \widetilde{x})$, $\widetilde{x} \in p^{-1}(x_0)$. Two coverings, $p_1\colon \widetilde{X}_1 \to X$ and $p_2\colon \widetilde{X}_2 \to X$, with connected and locally pathwise connected domains, are isomorphic if, and only if, the corresponding conjugate classes $\mathcal{H}_1(x_0)$ and $\mathcal{H}_2(x_0)$ are equal.

B. Under the same conditions, there exists a homomorphism $f\colon \widetilde{X}_1 \to \widetilde{X}_2$ if, and only if, every subgroup $H_1 \in \mathcal{H}_1(x_0)$ is contained in some subgroup $H_2 \in \mathcal{H}_2(x_0)$.

These two results show how the coverings can be classified by means of the subgroups of the fundamental group of the base space. It remains to establish an important complement, according to which, if X is semilocally simply connected, every conjugate class of subgroups in $\pi_1(X, x_0)$ is the class of some covering $p\colon \widetilde{X} \to X$. This will be proved later.

Example 7.11. What are all the coverings $p\colon \widetilde{X} \to S^1$, of the circle S^1, with $\widetilde{X}$ connected? If we identify the fundamental group of S^1 with $\mathbb{Z}$, their subgroups take the form $n\mathbb{Z}$, $n = 0, 1, 2, \ldots$. The coverings $p_n\colon S^1 \to S^1$, $p_n(z) = z^n$ determine the subgroups $n\mathbb{Z}$ with $n > 0$, while $p_0\colon R \to S^1$, $p_0(t) = e^{2\pi i t}$ determine the subgroup $\{0\}$. Any other covering $p\colon \widetilde{X} \to S^1$, with $\widetilde{X}$ connected, is isomorphic to one of these. Analogously, we prove that a covering of the real projective space P^n, with connected domain, must be a homeomorphism or is isomorphic to the covering of two leaves $p\colon S^n \to P^n$, since the fundamental group $\pi_1(P^n) = \mathbb{Z}_2$ has only two subgroups. ◁

7.4 Covering Automorphisms

The discussion of the previous section will now be restricted to the case of only one covering $p\colon \widetilde{X} \to X$.

We assume, in all of this section, that $\widetilde{X}$ is connected and locally pathwise connected.

Let $\widetilde{x}_0, \widetilde{x}_1 \in p^{-1}(x_0)$. As we proved above, there exists an endomorphism $f\colon \widetilde{X} \to \widetilde{X}$ such that $f(\widetilde{x}_0) = \widetilde{x}_1$ if, and only if, $H(\widetilde{x}_0) \subset H(\widetilde{x}_1)$. We also know that $H(\widetilde{x}_1) = \alpha^{-1}H(\widetilde{x}_0)\alpha$, where $\alpha \in \pi_1(X, x_0)$ is the homotopy class of $a = p \circ \widetilde{a}$ and $\widetilde{a}$ is a path in $\widetilde{X}$ that starts at $\widetilde{x}_0$ and ends at $\widetilde{x}_1$.

Example 7.12. Consider the nonregular covering $p\colon \widetilde{X} \to X$ presented in Example 7.2, and the points $\widetilde{x}_0, \widetilde{x}_1$ used in the same example. There does not exist an endomorphism $f\colon \widetilde{X} \to \widetilde{X}$ such that $f(\widetilde{x}_0) = \widetilde{x}_1$. In fact, if f existed, by letting $\widetilde{a}$ be the lifting of the path a from the point $\widetilde{x}_0$, $f \circ \widetilde{a}$ would be the lifting of a from the point $\widetilde{x}_1$. Now, we observe that, since $\widetilde{a}$ is closed, $f \circ \widetilde{a}$ also would be closed, but we saw in that example that the lifting of a starting at $\widetilde{x}_1$ is open. ◁

It also follows from the previous discussion that the endomorphism $f\colon \widetilde{X} \to \widetilde{X}$, with $f(\widetilde{x}_0) = \widetilde{x}_1$, is an automorphism if, and only if, $H(\widetilde{x}_0) = H(\widetilde{x}_1)$. Again, $H(\widetilde{x}_1) = \alpha^{-1}H(\widetilde{x}_0)\alpha$, $\alpha \in \pi_1(X, x_0)$ being the homotopy class of a path $a = p \circ \widetilde{a}$, where the path $\widetilde{a}$ starts at $\widetilde{x}_0$ and ends at $\widetilde{x}_1$, in the space $\widetilde{X}$.

If the covering $p\colon \widetilde{X} \to X$ is regular, it follows that, given any two points $\widetilde{x}_0, x_1 \in \widetilde{X}$ belonging to the same fiber $p^{-1}(x_0)$, there exists an endomorphism $f\colon \widetilde{X} \to \widetilde{X}$ such that $f(\widetilde{x}_0) = \widetilde{x}_1$. Besides this, every endomorphism of a regular covering is an automorphism. (See Corollary 7.12.)

In order to express the fact that, given any two points of the same fiber of a regular covering $p\colon \widetilde{X} \to X$, there exists an automorphism that maps one onto the other, we say that the automorphism group $G(\widetilde{X}|X)$ of a regular covering acts *transitively* in the fibers. This transitivity, besides being necessary, is also sufficient in order that $p\colon \widetilde{X} \to X$ be regular because it implies that if $\widetilde{a}$ is a lifting of the closed path a, the other liftings of a have the form $f \circ \widetilde{a}$, with $f \in G(\widetilde{X}|X)$; therefore, they are all open or all closed, according whether $\widetilde{a}$ is open or closed.

Another case where every endomorphism is an automorphism is the one of a covering $p\colon \widetilde{X} \to X$, with a finite number m of leaves. In fact, by Proposition 7.7, every endomorphism $f\colon \widetilde{X} \to \widetilde{X}$ is a covering, with a number n of leaves, necessarily finite. The equality $p \circ f = p$ implies that $p^{-1}(x) = f^{-1}(p^{-1}(x))$ for all $x \in X$, which gives us $m = nm$; hence, $n = 1$ and f is an automorphism.

Example 7.20 shows the existence of an endomorphism that is not an automorphism.

The *normalizer* of the subgroup H in a group G is the set $N(H)$ of all elements $g \in G$ such that $g^{-1}Hg = H$. The normalizer $N(H)$ is the largest subgroup of G that contains H as a normal subgroup. H is a normal subgroup of the group G if, and only if, $N(H) = G$.

In Proposition 7.13, it was established that, given the covering $p\colon \widetilde{X} \to X$, the fundamental group $\pi_1(X, x_0)$ acts transitively on the right in the fiber $p^{-1}(x_0)$, the isotropy group of $\widetilde{x} \in p^{-1}(x_0)$ being equal to $H(\widetilde{x})$. The action of an element $\alpha \in \pi_1(X, x_0)$ on $\widetilde{x} \in p^{-1}(x_0)$ is denoted by $\widetilde{x} \cdot \alpha$. We recall that $\widetilde{x} \cdot \alpha = \widetilde{a}(1)$ where $\widetilde{a}\colon I \to \widetilde{X}$ is the lifting, starting from $\widetilde{x}$, of a path $a\colon I \to X$ such that $\alpha = [a]$.

In terms of these notions, the existence of an endomorphism $f\colon \widetilde{X} \to \widetilde{X}$ with $f(\widetilde{x}_0) = \widetilde{x}_1$, where $\widetilde{x}_1 = \widetilde{x}_0 \cdot \alpha$, is equivalent to the statement that $H(\widetilde{x}_0) \subset \alpha^{-1} H(\widetilde{x}_0)\alpha$. Moreover, f is an automorphism if, and only if, $H(\widetilde{x}_0) = \alpha^{-1} H(\widetilde{x}_0)\alpha$; that is, if, and only if, $\alpha \in N(H(\widetilde{x}_0))$.

In particular, for each $\alpha \in N(H(\widetilde{x}_0))$, there exists a unique automorphism $f\colon \widetilde{X} \to \widetilde{X}$ such that $f(\widetilde{x}_0) = \widetilde{x}_0 \cdot \alpha$.

This is the crucial remark in order to prove the following proposition, which establishes an isomorphism between the group $G(\widetilde{X}|X)$ of the covering automorphisms $p\colon \widetilde{X} \to X$ and the quotient group $N(H(\widetilde{x}_0))/H(\widetilde{x}_0)$. In order to prove it, it is convenient to start with the following lemma.

Lemma 7.1. *Let $f\colon \widetilde{X} \to \widetilde{X}$ be an endomorphism of the covering $p\colon \widetilde{X} \to X$. For any $\widetilde{x} \in X$ and $\alpha \in \pi_1(X, p(\widetilde{x}))$, we have $f(\widetilde{x} \cdot \alpha) = f(\widetilde{x}) \cdot \alpha$.*

Proof. Let $\alpha = [a]$ and $\widetilde{a}$ be a lifting of a starting at the point $\widetilde{x}$. Then $\widetilde{x} \cdot \alpha = \widetilde{a}(1)$, so $f(\widetilde{x} \cdot \alpha) = f(\widetilde{a}(1))$. On the other hand, $f \circ \widetilde{a}$ is a lifting of a that starts at the point $f(\widetilde{x})$. Hence

$$f(\widetilde{x}) \cdot \alpha = (f \circ \widetilde{a})(1) = f(\widetilde{a}(1)) = f(\widetilde{x} \cdot \alpha),$$

which proves the lemma. □

Proposition 7.9. *Let $p\colon \widetilde{X} \to X$ be a covering, with $\widetilde{X}$ connected and locally pathwise connected. For each $\widetilde{x}_0 \in \widetilde{X}$, there exists a group isomorphism $G(\widetilde{X}|X) \approx N(H(\widetilde{x}_0))/H(\widetilde{x}_0)$.*

Proof. We define a map $\varphi\colon N(H(\widetilde{x}_0)) \to G(\widetilde{X}|X)$ by setting, for each $\alpha \in N(H(\widetilde{x}_0))$, $\varphi(\alpha) = f$, where $f\colon \widetilde{X} \to \widetilde{X}$ is the automorphism such that $f(\widetilde{x}_0) = \widetilde{x}_0 \cdot \alpha$. If $\varphi(\alpha) = f$ e $\varphi(\beta) = g$, we have $\widetilde{x}_0 \cdot \alpha = f(\widetilde{x}_0)$ and $\widetilde{x}_0 \cdot \beta = g(x_0)$. By the above lemma,

$$(f \circ g)(\widetilde{x}_0) = f(g(\widetilde{x}_0)) = f(\widetilde{x}_0 \cdot \beta) = f(x_0) \cdot \beta = x_0 \cdot \alpha\beta.$$

Therefore, $f \circ g = \varphi(\alpha\beta)$ and φ is a group homomorphism. We have $\varphi(\alpha) = id_{\widetilde{X}}$; that is, $\widetilde{x}_0 \cdot \alpha = \widetilde{x}_0$, if, and only if, $\alpha \in H(\widetilde{x}_0)$. Thus, $H(\widetilde{x}_0)$ is the kernel of φ. We claim that φ is surjective. In fact, given $f \in G(\widetilde{X}|X)$, let $f(\widetilde{x}_0) = \widetilde{x}_1$. Since $\pi_1(X, x_0)$ acts transitively on $p^{-1}(x_0)$, there exists $\alpha \in \pi_1(X, x_0)$ such that $\widetilde{x}_1 = \widetilde{x}_0 \cdot \alpha$. Since f is an automorphism, we have $\alpha \in N(H(\widetilde{x}_0))$. Since $f(\widetilde{x}_0) = \widetilde{x}_0 \cdot \alpha$, it follows that $f = \varphi(\alpha)$. The isomorphism theorem for groups yields, by passing to the quotient an isomorphism $\overline{\varphi}\colon N(H(\widetilde{x}_0))/H(\widetilde{x}_0) \to G(\widetilde{X}|X)$. □

Corollary 7.14. *If $\widetilde{X}$ is connected, locally pathwise connected, and the covering is regular, we have an isomorphism $G(\widetilde{X}|X) \approx \pi_1(X, x_0)/H(\widetilde{x}_0)$ for each $\widetilde{x}_0 \in p^{-1}(x_0)$.*

Corollary 7.15. *If $\widetilde{X}$ is simply connected, and locally pathwise connected then the group $G(\widetilde{X}|X)$ of automorphisms of the covering $p\colon \widetilde{X} \to X$ is isomorphic to the fundamental group $\pi_1(X, x_0)$.*

The isomorphism $\pi_1(X, x_0) \to G(\widetilde{X}|X)$ mentioned in Corollary 7.15 is defined by choosing a point $\widetilde{x}_0 \in \widetilde{X}$. It maps the element $\alpha \in \pi_1(X, x_0)$ into the automorphism $f\colon \widetilde{X} \to \widetilde{X}$, thus described: Given $\widetilde{x} \in \widetilde{X}$, we connect $\widetilde{x}$ to $\widetilde{x}_0$ by a path $\widetilde{b}$ in $\widetilde{X}$. Let $b = p \circ \widetilde{b}$, $a \in \alpha$ and $x = p(\widetilde{x})$. Then bab^{-1} is a closed path with base at the point x. The lifting of bab^{-1} from the point $\widetilde{x}$ ends at a point $\widetilde{y} \in p^{-1}(x)$. We set $f(\widetilde{x}) = \widetilde{y}$.

Example 7.13. **(Klein bottle)** Let G be the group generated by the homeomorphisms $f, g\colon \mathbb{R}^2 \to \mathbb{R}^2$, where $f(x, y) = (x, y+1)$ and $g(x, y) = (x+1, 1-y)$. Since $gf = f^{-1}g$, we can write the elements of G in the form $f^m \cdot g^n$, where m, $n \in \mathbb{Z}$. Now, $f^m g^n(x, y) = (x + n, y + m)$ if n is even and $f^m g^n(x, y) = (x + n, 1 - y + m)$ if n is odd. G is a properly discontinuous group of homeomorphisms of $\mathbb{R}^2$. In fact, given $z = (x, y)$, let V be the open square with center z, with sides of length 1, parallel to the axes. Then, for every $h \neq id$ in G, we have $V \cap hV \neq \varnothing$. The quotient space $K = \mathbb{R}^2/G$ is called the *Klein bottle*. The quotient map $p\colon \mathbb{R}^2 \to K$ exhibits the plane as a universal covering of the Klein bottle; therefore, the fundamental group of K is isomorphic to the group G of the automorphisms of this covering. Thus, the fundamental group of K has two generators, f, g, that satisfy the relation $gf = f^{-1}g$. (For other descriptions of the Klein bottle, see Examples 8.10 and 8.11, and Section 7 in this chapter.) ◁

Example 7.14. Now we give an example of a space whose fundamental group is $\mathbb{Z}_n$ (integers mod n). For this, we just have to consider a Hausdorff simply connected space Y which is also locally pathwise connected, and a group G de homeomorphisms of Y, isomorphic to $\mathbb{Z}_n$, such that none of them, with the exception of id_Y, has fixed points. (Being finite, G is properly discontinuous.) The quotient space Y/G will have fundamental group isomorphic to $\mathbb{Z}_n$. Let $D = \{z \in \mathbb{R}^2; |z| \leq 1\}$ be the unit disk of the plane and $X = \{(z,i); z \in D, i = 1,2,\ldots,n\}$ be the union of the n horizontal disjoint disks $D\times\{1\},\ldots,D\times\{n\}$. Y is the quotient space of X by the equivalence relation that identifies the points (z,i) and (z,j) when $|z| = 1$. If $n = 3$, Y is homeomorphic to the union of the sphere S^2 with its equatorial disk. For $n > 3$, we need $n-2$ curved "equatorial disks," all of them having in common the circle S^1. If $n = 2$, Y is homeomorphic to the sphere S^2. The space Y is simply connected, as we can see by induction, using Corollary 2.9. In order to define the homeomorphism $\varphi\colon Y \to Y$, we denote by $[z,i] \in Y$ the equivalence class of $(z,i) \in X$. Let $u = e^{2\pi i/n}$. Set $\varphi[z,i] = [u\cdot z, i+1]$ if $i < n$ and $\varphi[z,n] = [u\cdot z, 1]$. Note that φ does not have fixed points and the group generated by φ is $G = \{id_Y, \varphi, \ldots, \varphi^{n-1}\}$, isomorphic to $\mathbb{Z}_n$. Therefore, the quotient space Y/G has fundamental group isomorphic to $\mathbb{Z}_n$. When $n = 2$, φ is the antipodal map and Y/G is the real projective plane. ◁

Now we give an example, simpler than the previous one, of a space with fundamental group $\mathbb{Z}_n$, in dimension 3.

Example 7.15. Again, let $u = e^{2\pi i/n}$. The sphere S^3 is the set of pairs (z,w) of complex numbers such that $|z|^2 + |w|^2 = 1$. The homeomorphism $\varphi\colon S^3 \to S^3$, defined by $\varphi(z,w) = (u\cdot z, u\cdot w)$, generates the group $G = \{id, \varphi, \varphi^2, \ldots, \varphi^{n-1}\}$ of homeomorphisms without fixed points in S^3. G is a properly discontinuous group, isomorphic to $\mathbb{Z}_n$, so the quotient space $X = S^3/G$ has fundamental group isomorphic to $\mathbb{Z}_n$. When $n = 2$, X is the real projective space P^3. For any $n \in \mathbb{N}$, X is known as the *lens space* $L_{n,1}$. ◁

Example 7.16. Let X be the figure eight space, that is the union of two circles with a point in common. In Example 6.12 we showed a covering $p\colon \widetilde{X} \to X$. In that example, $\widetilde{X}$ is a tree; that is, a graph without cycles. We know that every tree is a contractible space. In particular, $\widetilde{X}$ is simply connected. Therefore, the fundamental group of X is isomorphic to the group of automorphisms $G(\widetilde{X}|X)$ of the covering $p\colon \widetilde{X} \to X$. Now, by taking the origin $\widetilde{x}_0$ as base point, an automorphism $f\colon \widetilde{X} \to \widetilde{X}$ is completely determined by the image $\widetilde{x}_1 = f(\widetilde{x}_0)$, which must be one of the "crossing points" of $\widetilde{X}$, and it can be any one of them. (Such points form the fiber over the point $x_0 \in X$, intersection of the two circles.) Now, the crossing points in $\widetilde{X}$ are in $1-1$ correspondence with the words $a^m b^n a^p \ldots$, because there exists a unique way to go from $\widetilde{x}_0$ to any other crossing point, along the segments of $\widetilde{X}$, without moving along the same segment twice. We conclude from this that the fundamental group of X is isomorphic to the free nonabelian group with two generators; each one of these generators is the homotopy class of one of the circles that form the space X. Considering a construction analogous to that of Example 12, Chapter 6, in three dimensions, we can show that the fundamental group of the union of three circles with one point in common is a free group with three generators. More generally, using the same method, it is possible to prove that if $X = X_1 \cup \ldots \cup X_n$ is the union of n circles with a point x_0 in common, then $\pi_1(X, x_0)$ is a free group with n generators. $\triangleleft$

7.5 Properly Discontinuous Groups and Regular Coverings

We have seen that if G is a properly discontinuous group of homeomorphisms of a topological space X, then the quotient map $\pi\colon X \to X/G$, onto the space of orbits of G, is a covering (Section 6.3) and when X is pathwise connected this covering is regular. (Example 6.3 of that chapter.) We now show that, conversely, every regular covering is essentially the quotient map onto the space of orbits of a properly discontinuous group of homeomorphisms.

Proposition 7.10. *Let $\widetilde{X}$ be a connected space. The automorphism group of a covering $p\colon \widetilde{X} \to X$ is a properly discontinuous group of homeomorphisms of the space $\widetilde{X}$.*

Proof. Given $\widetilde{x} \in \widetilde{X}$, let U be a distinguished neighborhood of $x = p(\widetilde{x})$ and V a neighborhood of $\widetilde{x}$ such that $p|V$ is a homeomorphism onto U. If the automorphism $f\colon \widetilde{X} \to \widetilde{X}$ is different from the identity, then $f(v) \neq v$ for every $v \in V$. Since v and $f(v)$ belong to the same fiber of the covering p and $p|V$ is injective, it follows that $f(v) \notin V$. Hence, $V \cap f(V) = \varnothing$, which proves that $G(\widetilde{X}|X)$ is properly discontinuous. $\square$

Given the covering $p\colon \widetilde{X} \to X$, with $\widetilde{X}$ connected, consider the covering $\pi\colon \widetilde{X} \to \widetilde{X}/G$, where $G = G(\widetilde{X}|X)$ is the properly discontinuous group whose elements are the automorphisms of the covering p.

Proposition 7.11. *Let X be a connected and locally pathwise connected space. If the covering $p\colon \widetilde{X} \to X$ is regular, there exists a homeomorphism $\xi\colon \widetilde{X}/G \to X$ which makes the diagram below commutative.*

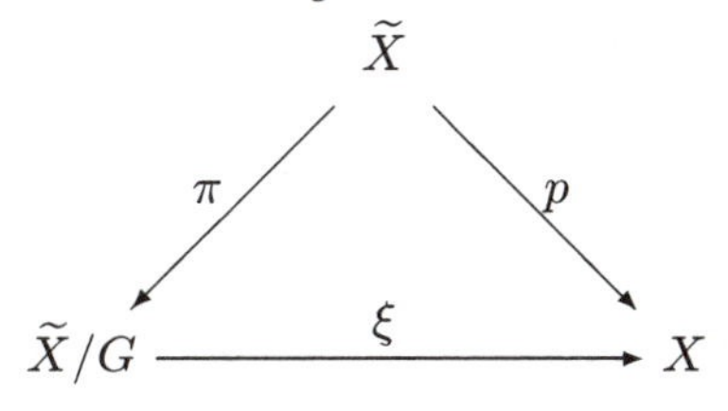

Proof. Given $\widetilde{x}$, $\widetilde{y} \in \widetilde{X}$, we have that $p(\widetilde{x}) = p(\widetilde{y}) \Leftrightarrow \widetilde{x}, \widetilde{y}$ belong to the same fiber of $p \Leftrightarrow$ there exists $f \in G$, such that $f(\widetilde{x}) = \widetilde{y}$ (because the covering is regular if, and only if, G is transitive in the fibers) $\Leftrightarrow G \cdot \widetilde{x} = G \cdot \widetilde{y} \Leftrightarrow \pi(\widetilde{x}) = \pi(\widetilde{y})$. Hence, the equivalence relations determined by p and π in $\widetilde{X}$ coincide. This gives us a continuous bijection $\xi\colon \widetilde{X}/G \to X$ such that $\xi \circ \pi = p$. Since p is open, the same happens with ξ. Hence, ξ is a homeomorphism. □

7.6 Existence of Coverings

We start with the following question: Which topological spaces have a simply connected covering? It is easy to obtain a necessary condition. If $p\colon \widetilde{X} \to X$ is a covering with $\widetilde{X}$ simply connected, then X must be semi-locally simply connected. In fact, if $V \subset X$ is a distinguished neighborhood, then every closed path a, contained in V, has a closed lifting $\widetilde{a}$. Since $\widetilde{X}$ is simply connected, $\widetilde{a}$ is homotopic to a constant in $\widetilde{X}$ and from this we conclude that $a = p \circ \widetilde{a}$ is also homotopic to a constant in X.

We show below (for locally pathwise connected spaces), that this condition is also sufficient for the existence of a "universal" covering $p\colon \widetilde{X} \to X$; that is, a covering with $\widetilde{X}$ simply connected. Moreover, we prove that if X is locally pathwise connected and semi-locally simply connected then, for every subgroup $H \subset \pi_1(X, x_0)$ there exists a covering $p\colon \widetilde{X} \to X$ such that $p_{\#}\pi_1(\widetilde{X}, \widetilde{x}_0) = H$. This will complete the discussion we started at the end of Section 3. Given a connected, locally pathwise connected and semi-locally simply connected space X, we fix $x_0 \in X$. To each covering $p\colon \widetilde{X} \to X$, with $\widetilde{X}$ connected, corresponds a conjugate class $\mathcal{H}(x_0)$ of subgroups of $\pi_1(X, x_0)$. Two connected coverings with base X are isomorphic if, and only if, to them corresponds the same conjugate class. Now we see that such correspondence between classes of isomorphic coverings with base X and conjugate classes of subgroups of $\pi_1(X, x_0)$ is surjective and therefore bijective, under these topological hypothesis on X.

Proposition 7.12. *Let X be a connected, locally pathwise connected, and semi-locally simply connected space. Given $x_0 \in X$ and a subgroup $H \subset \pi_1(X, x_0)$, there exists a covering $p\colon \widetilde{X} \to X$, with $\widetilde{X}$ connected, and a point $\widetilde{x}_0 \in \widetilde{X}$ such that $p_{\#}\pi_1(\widetilde{X}, \widetilde{x}_0) = H$.*

Proof. Let a, b be two paths in X, starting at the point x_0. We say that a and b are equivalent, and we write $a \equiv b$, when $a(1) = b(1)$ and $[ab^{-1}] \in H$. We use the notation $\langle a \rangle$ to represent the equivalence class of the path a. Let $\widetilde{X}$ be the set of all equivalence classes $\langle a \rangle$ of the paths a in X that start at the point x_0. We define a map $p\colon \widetilde{X} \to X$ by setting $p(\langle a \rangle) = a(1)$. In order to introduce a topology in $\widetilde{X}$, we consider the basis of X formed by all pathwise connected open sets $U \subset X$ such that every path in U is homotopic to a constant in X. For each point $\langle a \rangle \in \widetilde{X}$ and each open set $U \in \mathcal{U}$ such that $a(1) \in U$, we set

$$\widetilde{U}\langle a \rangle = \{\langle ab \rangle; b(I) \subset U\}.$$

The sets $\widetilde{U}\langle a \rangle$ form the basis for a topology in $\widetilde{X}$, according to which $p\colon \widetilde{X} \to X$ is continuous and open. Besides this, $p|\widetilde{U}\langle a \rangle$ is a bijection (and therefore, a homeomorphism) onto U. For each $U \in \mathcal{U}$, the inverse image $p^{-1}(U)$ is the union of the sets $\widetilde{U}\langle a \rangle$, where a varies among the paths in X with origin x_0 and final point in U. Two of these sets either coincide or are disjoint. It follows that $p\colon \widetilde{X} \to X$ is a covering. Given a path $a\colon I \to X$, with origin x_0, for each $t \in I$ let $a_t\colon I \to X$ be the path given by $a_t(s) = a(st)$. Then $\widetilde{a}\colon I \to \widetilde{X}$, defined by $\widetilde{a}(t) = \langle a_t \rangle$, is the lifting of a starting at the point $\widetilde{x}_0 = \langle e_{x_0} \rangle$, where e_{x_0} is the constant path in X, at the point x_0. Note that every point $\langle a \rangle \in \widetilde{X}$ can be connected to the point $\widetilde{x}_0$ by the path $t \mapsto \langle a_t \rangle$. Hence, $\widetilde{X}$ is pathwise connected. We have $[a] \in p_{\#}\pi_1(\widetilde{X}, \widetilde{x}_0) \Leftrightarrow \widetilde{a}$ is a closed path $\Leftrightarrow \langle a \rangle = \langle a_1 \rangle = \langle e_{x_0} \rangle \Leftrightarrow [a] \in H$. This completes the proof. □

Corollary 7.16. *Every connected, locally pathwise connected, and semi-locally simply connected topological space X admits a covering $p\colon \widetilde{X} \to X$, with $\widetilde{X}$ simply connected. Any two of these coverings are isomorphic.*

Example 7.17. By Corollary 7.16, the group $\mathrm{SO}(n)$ has a covering $p\colon \widetilde{X} \to \mathrm{SO}(n)$, with $\widetilde{X}$ simply connected. Since $\pi_1(\mathrm{SO})(n) = \mathbb{Z}_2$, this covering has two leaves. (See Corollary 7.7.) By Proposition 7.5, there exists a group structure in $\widetilde{X}$ which turns the projection $p\colon \widetilde{X} \to \mathrm{SO}(n)$ into a homomorphism. This group $\widetilde{X}$ is called the *group of spinors of order n* and it is denoted by $\mathrm{Spin}(n)$. We have $\mathrm{Spin}(3) = S^3$, and $\mathrm{Spin}(4) = S^3 \times S^3$. The group structure in $\mathrm{Spin}(n)$, obtained by means of the requirement that $p\colon \mathrm{Spin}(n) \to \mathrm{SO}(n)$ be an homomorphism, is unique, provided that we choose one of the two elements that are mapped by p in the identity matrix to be the neutral element. ◁

7.7 Fundamental Group of a Compact Surface

In order to finish this chapter, we shall determine, in terms of generators and relations, the fundamental group of a compact surface. By "surface,"

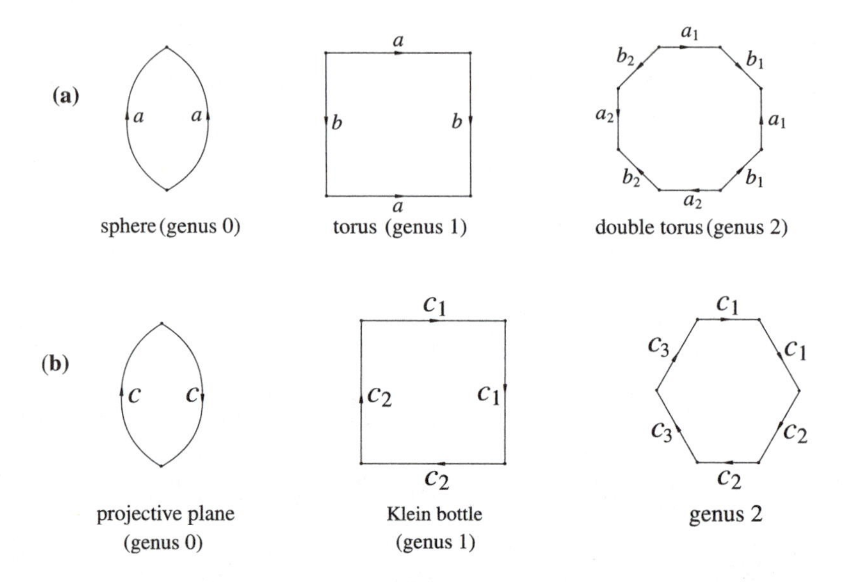

Figure 7.2. Some identification schemes for surface construction.

we mean a topological manifold of dimension 2; that is, a Hausdorff topological space that is locally homeomorphic to the Euclidean plane $\mathbb{R}^2$.

It is proven in combinatorial topology (see Seifert & Threlfall (1980)) that every compact surface is the quotient space of a plane polygon by an equivalence relation according to which the sides that make up the boundary of the polygon are identified two by two, according to schemes such as those illustrated in Figure 7.2.

There are three identifying schemes. The first is the one of the *orientable surface of genus zero*, which is homeomorphic to the sphere S^2, in which the "polygon" has two sides, which must be glued one to the other, as indicated in the leftmost picture in Figure 7.2(a).

The second type of scheme is that of an *orientable surface of genus* $g \geq 1$. The polygon has $4g$ sides, labeled $a_1, b_1, a_1, b_1, \ldots, a_g, b_g$. Each of these sides is oriented by means of an arrow. Moving along the boundary of the polygon in the clockwise sense, the directions of the arrows provides a "word" $\omega = a_1 b_1 a_1^{-1} b_1^{-1} a_2 b_2 a_2^{-1} b_2^{-1} \ldots a_g^{-1} b_g^{-1}$, which represents a closed path in the surface. This is again illustrated for $g = 1$ and $g = 2$ in Figure 7.2(a). (middle and right pictures). The picture on the left in Figure 7.3 illustrates the general scheme.

The third type of identifying scheme is that of a *nonorientable surface of genus* $g = h - 1$. The polygon has $2h$ sides, labeled $c_1, c_1, c_2, c_2, \ldots, c_h, c_h$. These sides are oriented by arrows in such a way that, by moving

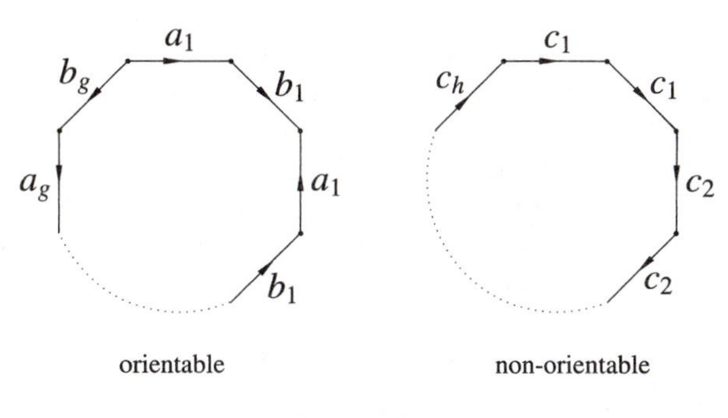

Figure 7.3.

along the boundary of the polygon in the clockwise sense, the arrows point in our direction. This gives us the word $\lambda = c_1^2 c_2^2 \dots c_h^2$, which represents a closed path in the surface. This is illustrated for $g = 0$, $g = 1$ and $g = 2$ in Figure 7.2(b). The picture on the right in Figure 7.3 illustrates the general scheme.

Note that, both the image (by the quotient map) of the path ω of the second scheme as well as the one of the path λ of the third scheme are homotopic to constant paths in the corresponding surfaces, because ω and λ are homotopic to a constant on their polygons.

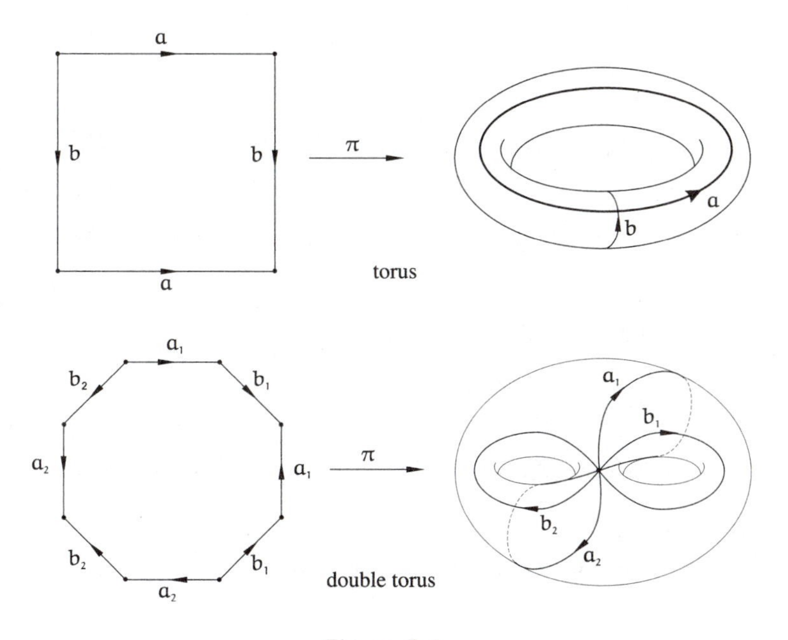

Figure 7.4.

If $\pi\colon P \to S$ is the quotient map of the polygon P onto the surface S, the boundary of P is transformed by π into a union of circles with a point in common. The number of circles is $2g$ for an orientable surface of genus $g \geq 1$ and $h = g + 1$ for a nonorientable surface of genus $g \geq 1$. The interior of the polygon P is mapped by π homeomorphically onto the complementary set of this union of circles on the surface. This is illustrated in Figure 7.4 for the orientable surfaces of genus 1 (torus) and genus 2 (double torus).

Proposition 7.13. *The fundamental group of a compact orientable surface of genus $g \geq 1$ has $2g$ generators $\alpha_1, \beta_1, \alpha_2, \beta_2, \dots, \alpha_g, \beta_g$ and only one relation,*

$$\alpha_1\beta_1\alpha_1^{-1}\beta_1^{-1}\alpha_2\beta_2\alpha_2^{-1}\beta_2^{-1}\dots\alpha_g\beta_g\alpha_g^{-1}\beta_g^{-1} = 1.$$

The fundamental group of a compact nonorientable surface of genus g has $h = g + 1$ generators $\gamma_1, \gamma_2, \dots, \gamma_h$ and only one relation,

$$\gamma_1^2\gamma_2^2\dots\gamma_h^2 = 1.$$

The above proposition is a particular case of the following situation: we have a space X, a closed subset $A \subset X$, and a continuous map $f\colon D \to X$, of the unit disk $D = \{(x, y) \in \mathbb{R}^2; x^2 + y^2 \leq 1\}$ on X, satisfying the following conditions:

1. $f(S^1) \subset A$;

2. $f|\operatorname{int.} D$ is a homeomorphism onto $X - A$.

In this case, we say that X is obtained from A by the *adjunction of a two-dimensional cell.* Let $e = f(D)$, $a = f|S^1$. We then write $X = A \bigcup_a e$. By choosing base points $u_o \in S^1$ and $x_0 = f(u_0) \in A$, the homotopy class $[a]$ is an element of the fundamental group $\pi_1(A, x_0)$. When X is a compact surface, the subset A is the union of a finite number of circles with a point in common and f is the quotient map.

More generally, suppose that we have a family of continuous maps $a_\lambda\colon S^1 \to A$, $\lambda \in L$, all of them taking values in a topological space A, and consider the disjoint union $Z = A \cup (\bigcup_{\lambda \in L} D_\lambda)$, where each D_λ is a copy of the disk D. We consider in Z the "sum" topology, in which A and each D_λ are open and closed. We introduce in Z the equivalence relation that identifies each point $z \in S_\lambda^1$ (boundary of D_λ) with its image $a_\lambda(z) \in A$. Let X be the quotient space of Z by this equivalence relation. We set

$$X = A \bigcup_{a_\lambda} \{e_\lambda\},$$

where e_λ is the image of D_λ by the quotient map $f\colon Z \to X$. Let $f_\lambda = f|D_\lambda$ and $\dot{e}_\lambda = f_\lambda(S^1_\lambda) = a_\lambda(S^1_\lambda) = e_\lambda \cap A$. We say that X *is obtained from* A *by the adjunction of the 2-dimensional cells* e_λ. The following properties hold:

1. For each $\lambda \in L$, f_λ is a homeomorphism of the interior of the disk D_λ onto the open set $e_\lambda - \dot{e}_\lambda$ of the space X;

2. If $\lambda \neq \mu$ then $e_\lambda - \dot{e}_\lambda$ and $e_\mu - \dot{e}_\mu$ are disjoint;

3. The set $S \subset X$ is closed (respectively open) if, and only if, for each $\lambda \in L$, the intersection $S \cap e_\lambda$ is closed (respectively open) in e_λ.

In the proposition below, which contains Proposition 7.13, we suppose that A is pathwise connected, so that the fundamental group $\pi(A, x_0)$ does not depend essentially of the base point x_0.

Proposition 7.14. *Given a pathwise connected and semi-locally simply connected space* A*, let* $X = A \bigcup_a e$ *be the space obtained from* A *by the adjunction of the 2-dimensional cell* e*, by means of the continuous map* $a\colon S^1 \to A$*. By setting* $x_0 = a(u_0)$, $u_0 \in S^1$*, the fundamental group* $\pi_1(X, x_0)$ *is isomorphic to the quotient group of* $\pi_1(A, x_0)$ *by the normal subgroup generated by* $[a]$.

Proof. Let $j\colon A \to X$ be the inclusion map. We must show:

A. The induced homomorphism $j_\#\colon \pi_1(A, x_0) \to \pi_1(X, x_0)$ is surjective;

B. The kernel of $j_\#$ is the normal subgroup generated by $[a]$.

Proof of A. Let $y \in X - A$ (for example, the center of the cell e). Set $U = X - A$ and $V = X - \{y\}$. Then U and V are open sets in X, $U \cap V$ is homeomorphic to the open disk minus a point (and therefore it is pathwise connected) and $X = U \cup V$. It follows from Proposition 2.11 that $\pi_1(X, x_0)$ is generated by the images of $\pi_1(U)$ and $\pi_1(V)$, induced by the inclusions $U \to X$ and $V \to X$. Now, U is contractible because it is homeomorphic to an open disk, and the inclusion $A \to V$ is a homotopy equivalence: Its homotopic inverse $V \to A$ is the retraction that projects each point of $V - A$ radially from the point y and leave fixed the points of A. Hence, $\pi_1(U)$ is trivial and the homomorphisms $\pi_1(A, x_0) \to \pi_1(V, x_0)$ and $\pi_1(V, x_0) \to \pi_1(X, x_0)$, induced by inclusions, are surjective; hence, the composite homomorphism $j_\#$ is also surjective.

Proof of B. Let N be the normal subgroup of $\pi_1(A, x_0)$ generated by the class $[a]$. By Proposition 7.12, there exists a covering $p\colon \widetilde{A} \to A$ such that $p_\#\pi_1(\widetilde{A}, \widetilde{x}_0) = N$, for all $\widetilde{x}_0 \in p^{-1}(x_0)$. Since $[a] \in N$, the map $a\colon S^1 \to A$

has a lifting $\widetilde{a}_\lambda\colon S^1 \to \widetilde{A}$, with $\widetilde{a}_\lambda(u_0) = \lambda$, for each $\lambda \in p^{-1}(x_0)$. Let $\widetilde{X}$ be the space obtained from $\widetilde{A}$ by adjunction of two-dimensional cells through the maps $\widetilde{a}_\lambda$. There exists a covering map $q\colon \widetilde{X} \to X$ such that $q|\widetilde{A} = p$. (In order to see this, note that the covering $p\colon \widetilde{A} \to A$ is regular; hence, $A = \widetilde{A}/G$, where $G = G(\widetilde{A}|A)$. Now, every $g \in G$ extends in an evident way to a homeomorphism of $\widetilde{X}$ and G becomes a properly discontinuous group in $\widetilde{X}$, with $\widetilde{X}/G = X$.) Now we take, arbitrarily, a closed path b in A, with base at the point x_0, such that b is homotopic to a constant path in X (this means that $[b] \in \pi_1(A, x_0)$ is in the kernel of $j_\#$). Then the lifting $\widetilde{b}$ of the path b, from any point $\lambda \in p^{-1}(x_0)$, is a closed path. But $\widetilde{b}$ is a path in $\widetilde{A}$. Since it is closed, this means that $[b] \in N$. (See Corollary 7.1.) Thus, the kernel of the homomorphism $j_\#\colon \pi_1(A, x_0) \to \pi_1(X, x_0)$ is contained in the normal subgroup N, generated by $[a]$. But it is obvious that $[a]$ belongs to the kernel of $j_\#$, which is a normal subgroup; hence, such a kernel contains N. This concludes the proof. □

Proposition 7.13 follows from Proposition 7.14 by virtue of Example 7.16, according to which the fundamental group of the space that consists of n circles with a point in common is a free subgroup with n generators.

Remark. With the exception of the sphere and the projective plane, the universal covering space of a compact surface is the plane $\mathbb{R}^2$. This follows from the fact that the fundamental groups of these surfaces are infinite; hence, their universal covering spaces are not compact. Since the covering map is a local homeomorphism, each covering space of a surface is also a surface. But, the only noncompact simply connected surface is the plane $\mathbb{R}^2$. The proof of this fact is surprisingly nontrivial. (See Seifert & Threlfall (1980), page 332. Another proof can be given as a consequence of the Koebe uniformization theorem for Riemann surfaces.)

Example 7.18. Completing the discussion that we started in Example 14 of Chapter 6, we see that, with the exception of the torus, no compact surface admits the structure of a topological group. In fact, the fundamental group of a compact surface is nonabelian, with three exceptions: the sphere, the projective plane and the torus. We have already seen (in the example mentioned above) that the sphere is not a topological group. By Proposition 7.5, it follows that the projective plane also is not a topological group, because it is covered by the sphere. The torus is a group, so all the cases are now covered. ◁

Example 7.19. As a consequence of Proposition 7.13, the fundamental group of the Klein bottle has two generators, say c, d, that satisfy the unique relation $c^2d^2 = 1$, which is equivalent to $cd = c^{-1}d^{-1}$. By setting $a = c$ and $b = dc$, we see that the same group admits the generators a, b, which satisfy the unique relation $ab = b^{-1}a$ (because $ab = cdc = c^{-1}d^{-1}c = b^{-1}a$). This takes us back to the description of the same group given in Example 7.12 and illustrates the fact that the presentation of a group by using generators and relations can, in general, be done in different ways, apparently distinct. ◁

Example 7.20. By virtue of Proposition 7.8 and its Corollary 7.12, one obtains an endomorphism $f\colon \widetilde{X} \to \widetilde{X}$ which is not an automorphism when the covering $p\colon \widetilde{X} \to X$ has the following property: there exist $x_0 \in X$, $\widetilde{x}_0 \in p^{-1}(x_0)$, and $\alpha \in \pi_1(X, x_0)$ such that $\alpha^{-1} \cdot H(\widetilde{x}_0) \cdot \alpha$ is a proper subgroup of $H(\widetilde{x_0})$. With this purpose, we take X as the space obtained by the adjunction of a two-dimensional cell to the figure eight space through the map $aba^{-1}b^{-2}\colon S^1 \to$ figure eight, where a and b are the two canonical closed paths in the figure eight space. By Proposition 7.14, the fundamental group of X (with base in x_0, crossing point in the figure eight) has the generators $[a] = \alpha$ and $[b] = \beta$, with the unique relation $\alpha\beta\alpha^{-1} = \beta^2$. By Proposition 7.12, there exists a covering $p\colon (\widetilde{X}, \widetilde{x}_0) \to (X, x_0)$ such that $H(\widetilde{x}_0) = p_{\#}\pi_1(\widetilde{X}, x_0)$ is the cyclic subgroup of $\pi_1(X, x_0)$ generated by β. Then $\alpha^{-1}{\cdot}H(\widetilde{x}_0){\cdot}\alpha = H(\widetilde{x}_1)$, where $\widetilde{x}_1 = \widetilde{x}_0{\cdot}\alpha$. It follows that $H(\widetilde{x}_1)$ is the proper subgroup of $H(\widetilde{x}_0)$ whose elements are the powers of β with an even exponent. Therefore, there exists an endomorphism $f\colon (\widetilde{X}, \widetilde{x}_1) \to (\widetilde{X}, \widetilde{x}_0)$, which is not an automorphism. ◁

7.8 Exercises

1. Let X be the figure eight space and $\widetilde{X}$ the subset of the upper half-plane, formed by the horizontal axis, along with the circles of radius $1/3$, tangent to this axis at the points $(n, 0)$, $n \in Z$. Define a covering map of $\widetilde{X}$ onto X. Determine whether this covering is regular or not. Take $x_0 = (0, 0)$, describe the conjugate class of $\pi_1(X, x_0)$ defined by this covering, and determine $G(\widetilde{X}|X)$.

2. In Exercise 3 of Chapter 6, show that p regular implies q regular.

3. Let X be an arbitrary topological space. Given the covering $\xi\colon \mathbb{R} \to S^1$, $\xi(t) = e^{it}$, prove that a continuous map $f\colon X \to S^1$ has a lifting with respect to ξ if, and only if, it is homotopic to a constant.

4. If $n \geq 2$, then every continuous map $f\colon P^n \to S^1$ is homotopic to a constant.

5. Every continuous map from the sphere S^2 to the torus T^2 is homotopic to a constant. The same occurs with maps from S^2 to S^1.

6. Let M be a compact orientable surface of genus $g > 1$. Show that there exists $f: M \to S^1$ continuous, nonhomotopic to a constant.

7. Find a covering of the figure eight space such that $H(\widetilde{x}_0)$ is a cyclic group. Is this covering regular? What is the group $G(\widetilde{X}|X)$ of the automorphisms of this covering?

8. In Exercise 15 of Chapter 6, if $p: \widetilde{X} \to X$ is a regular covering of n leaves, and $f = p$, show that $\widetilde{Z}$ is the union of n disjoint copies of $\widetilde{X}$.

9. Let X be the union of two circles tangent at the point x_0 and $\widetilde{X}$ be the grid consisting of the points of the plane that have at least one integral coordinate. Define a covering map $p: \widetilde{X} \to X$ such that, by fixing a point $\widetilde{x}_0 \in p^{-1}(x_0)$, $H(\widetilde{x}_0)$ is the commutator subgroup (generated by the elements of the form $\alpha\beta\alpha^{-1}\beta^{-1}$) in $\pi_1(X, x_0)$ and $G(\widetilde{X}|X) = \mathbb{Z} \oplus \mathbb{Z}$.

10. Determine all of the connected coverings of the torus $T^n = S^1 \times \ldots \times S^1$.

11. Let $p: \widetilde{X} \to X$ be a covering with $\widetilde{X}$ connected, $G(\widetilde{X}|X)$ be the automorphism group and $\pi: \widetilde{X} \to \widetilde{X}/G(\widetilde{X}|X)$ the quotient map. There exists a continuous map $q: \widetilde{X}/G(\widetilde{X}|X) \to X$ such that $q \circ \pi = p$. Both π and q are covering maps.

12. Let $X = S^1 \cup S^2$ be the union of a circle and a sphere, with $S^1 \cap S^2 = \{x_0\}$. Obtain the universal covering of X.

13. What is the universal covering of the space formed by the union of a torus with a circle that has a point in common with it?

14. Let $p: \widetilde{G} \to G$ be a homomorphic covering. If $\widetilde{G}$ is simply connected, then $\pi_1(G)$ is isomorphic to the kernel of p.

15. Restate, in terms of the complex integral

$$\int_a \frac{f'(z)}{f(z)} dz,$$

the conditions of the Examples 7.4 and 7.5, in order that the holomorphic function $f: U \to \mathbb{C} - \{0\}$ have, respectively, a branch of $\log f(z)$ and a branch of $\sqrt[k]{f(z)}$ defined globally in U.

16. We say that a topological space is triangulable when it is homeomorphic to a polyhedron. Let $p: \widetilde{X} \to X$ be a covering. If X if triangulable, prove that $\widetilde{X}$ is also triangulable.

17. Let α_i be the number of simplices of dimension i of the polyhedron P. If X is a space homeomorphic to P, the number

$$\chi(X) = \sum_{i\geq 0}(-1)^i \alpha_i$$

is called the *Euler characteristic* of X. If $p\colon \widetilde{X} \to X$ is a covering with n leaves of the triangulable space X, prove that $\chi(\widetilde{X}) = n\cdot\chi(X)$. By taking into account that $\chi(S^{2n}) = 2$, show that a covering $p\colon S^{2n} \to X$ has at most two leaves. Therefore, if the universal covering of X is S^{2n}, then X is homeomorphic to S^{2n} or $\pi_1(X) = \mathbb{Z}_2$. Conclude also that the only finite group that acts freely on S^{2n} is $\mathbb{Z}_2$.

18. Let $p\colon \widetilde{M} \to M$ be a differentiable covering (by this we mean, in particular, that p is a local diffeomorphism between the manifolds $\widetilde{M}$ and M). If M admits a continuous non-null tangent vector field, the same happens to $\widetilde{M}$.

19. Use the first part of Exercise 15 to prove the converse of Exercise 16 in the case where $\widetilde{M}$ is compact. (Admit the theorem from differential topology that relates $\chi(M)$ with the existence of continuous and non-null tangent vector fields to M.)

20. Let ω be a closed differential form of degree 1 and class C^∞ in the manifold M and x_0 a point of M. Consider the subgroup $H \subset \pi_1(M, x_0)$ whose elements are the homotopy classes $\alpha = [a]$ of the paths $a\colon (I, \partial I) \to (M, x_0)$ such that $\int_a \omega = 0$ and the covering $p\colon \widetilde{M} \to M$ such that $p_\# \pi_1(M, \widetilde{x}_0) = H$ for some $\widetilde{x}_0 \in p^{-1}(x_0)$. Prove that there exists a function $f\colon \widetilde{M} \to \mathbb{R}$, of class C^∞, such that $p^*\omega = df$.

21. In Exercise 20, show that the covering $p\colon \widetilde{M} \to M$ is regular and that its automorphism group $G(\widetilde{M}|M)$ is isomorphic to the group of "periods" of ω; that is, the additive group of real numbers $\int_a \omega$, where $[a] \in \pi_1(M, x_0)$.

22. Let Z be the simply connected (but not locally connected) space defined in Exercise 16 of Chapter 2, and $Y \subset Z$ the arc there mentioned. Denote by X' the interval $[0, 1/\pi]$ of the horizontal axis and let $W = X' \cup Y$. Define a continuous bijection $\varphi\colon Z \to W$ by setting $\varphi(x, sen(1/x)) = x$ if $x \in [0, 1/\pi]$, and $\varphi(x, y) = (x, y)$ if $(x, y) \in Y$. Show that $\varphi^{-1}\colon W \to Z$ is not continuous. Consider a homeomorphism $h\colon W \to S^1$ and show that the continuous bijection $g = h \circ \varphi\colon Z \to S^1$ does not have a continuous lifting $\widetilde{g}\colon Z \to \mathbb{R}$ with respect to the universal covering $\xi\colon \mathbb{R} \to S^1$. Conclude that $g\colon Z \to S^1$ is not homotopic to a constant. (See Exercise 2, above.)

Chapter 8

Oriented Double Covering

In this chapter, we study an example of covering that has applications to topology and geometry. Since we will treat this topic in detail, we decided to include it as a separate chapter, which justifies itself because there are few comprehensive expositions of this subject in the literature.

8.1 Orientation of a Vector Space

Let E be a vector space of dimension m over the field of real numbers.

A *basis* in E is an ordered list $\mathcal{E} = (e_1, \dots, e_m)$ of m linearly independent vectors. If $\mathcal{F} = (f_1, \dots, f_m)$ is another basis in E, there exists a unique $m \times m$ invertible real matrix, $A = (a_{ij})$ such that

$$f_j = \sum_{i=1}^{m} a_{ij} e_i$$

for every $j = 1, 2, \dots, m$. A is called the *transition matrix* from the basis $\mathcal{E}$ to the basis $\mathcal{F}$.

Given two bases $\mathcal{E}$ and $\mathcal{F}$ in E, we say that $\mathcal{E}$ and $\mathcal{F}$ are *equally oriented*, and we denote this by $\mathcal{E} \equiv \mathcal{F}$, when the transition matrix from $\mathcal{E}$ to $\mathcal{F}$ has a positive determinant. The relation $\mathcal{E} \equiv \mathcal{F}$ is an equivalence relation in the set of bases of the space E. Since a transition matrix has either a positive or a negative determinant, this equivalence relation has precisely two equivalence classes.

Each one of these equivalence classes is called an *orientation* of the vector space E.

Thus, an orientation in the vector space E is a set $\mathcal{O}$ of bases of E with the following property: if a basis $\mathcal{E}$ belongs to $\mathcal{O}$ and $\mathcal{F}$ is any basis in E, then $\mathcal{F} \in \mathcal{O}$ if, and only if, the transition matrix from $\mathcal{E}$ to $\mathcal{F}$ has a positive determinant.

Given an orientation $\mathcal{O}$ in the vector space E, the other orientation of E is called the *opposite orientation* of $\mathcal{O}$ and it will be denoted by $-\mathcal{O}$.

Every basis $\mathcal{E}$ in E defines an orientation $\mathcal{O}$ in the space E. $\mathcal{O}$ consists of the bases of E that are equally oriented with respect to the basis $\mathcal{E}$.

In order that two bases $\mathcal{E} = (e_1, \ldots, e_m)$ and $\mathcal{F} = (f_1, \ldots, f_m)$ be equally oriented it is necessary and sufficient that there exist m paths $h_j \colon I \to E$ such that, for every $t \in I$ the list $\mathcal{H}(t) = (h_1(t), \ldots, h_m(t))$ is a basis of E, with $\mathcal{H}(0) = \mathcal{E}$ and $\mathcal{H}(1) = \mathcal{F}$. To prove this, we should recall that the set of $m \times m$ matrices with positive determinant is pathwise connected. If we have

$$f_j = \sum_{i=1}^{m} a_{ij} e_i, \quad j = 1, \ldots, m, \quad \text{with} \quad \det(a_{ij}) > 0,$$

then we take a matrix path $A(t) = (a_{ij}(t))$, satisfying $\det A(t) > 0$ for all $t \in I$, $A(0) = m \times m$ identity matrix, $A(1) = (a_{ij})$, and we set

$$h_j(t) = \sum_{i=1}^{m} a_{ij}(t) e_i.$$

Thus, two bases of E are equally oriented if, and only if, one of them can be continuously deformed onto the other, in such a way that at each instant of the deformation we have a basis of E. (For topological facts in E, we take an arbitrary norm of the space. It is well known that all norms in a finite dimensional vector space define the same topology.)

An *oriented vector space* is a pair $(E, \mathcal{O})$, where E is a vector space of finite dimension over the field of real numbers and $\mathcal{O}$ is an orientation of E. It is very common to refer to such a space by using only the vector space notation E; that is, the orientation does not appear explicitly.

The vector space $\mathbb{R}^m$ will always be considered with the orientation defined by the canonical basis $\mathcal{E} = (e_1, \ldots, e_m)$, where $e_1 = (1, 0, \ldots, 0)$, $e_2 = (0, 1, 0, \ldots, 0)$.

In an oriented vector space $(E, \mathcal{O})$, the bases that belong to $\mathcal{O}$ are called *positive* and the other bases are called *negative.*

Let E, F be oriented vector spaces with the same dimension m. An isomorphism $f \colon E \to F$ is said to be *positive* when it transforms any positive basis of E onto a positive basis of F. In order that this occur, it suffices for f to transform *one* positive basis of E onto a positive basis of F. When an isomorphism from E onto F is not positive, we say that it is *negative.*

A linear transformation $T \colon \mathbb{R}^m \to \mathbb{R}^m$ is a positive isomorphism if, and only if, its matrix with respect to the canonical basis of $\mathbb{R}^m$ has determinant > 0.

When an isomorphism $f\colon E \to F$, between two oriented vector spaces, is positive, we say that f is *orientation preserving.* If only one the these vector spaces is oriented, the requirement that the isomorphism $f\colon E \to F$ be positive determines in a unique way an orientation in the other space.

8.2 Orientable Manifolds

By "manifold," we mean differentiable manifold (Hausdorff with countable basis), of a certain class C^k which we specify when it is necessary. The differentiable maps will belong to this class C^k.

Let M, N be manifolds of the same dimension. For each $x \in M$ and each $y \in N$ we choose, arbitrarily, an orientation $\mathcal{O}_x$ in T_xM (the vector space tangent to M at the point x) and an orientation $\mathcal{O}'_y$ in T_yN. Let $f\colon M \to N$ be a local diffeomorphism. We say that f is *positive* (with respect to the chosen orientations) when, for each $x \in M$, the linear isomorphism $f'(x)\colon T_xM \to T_{f(x)}N$ is positive. Analogously, we define a *negative* local diffeomorphism; in this case, we must require that, for all $x \in M$, the linear isomorphism $f'(x)\colon T_xM \to T_{f(x)}N$ reverses the orientation. We must observe that there may exist local diffeomorphisms which are neither positive nor negative.

Evidently, it is not interesting to choose, in a random way, an orientation in each tangent vector space of a manifold without any correlation with one another. We impose now that this choice be, in a certain sense, continuous.

An *orientation* $\mathcal{O}$ in a differentiable manifold M is a correspondence that associates to each point $x \in M$ an orientation $\mathcal{O}_x$ in the tangent vector space T_xM, in such a way that every point $x \in M$ belongs to the domain U of a positive coordinate system $\varphi\colon U \to \mathbb{R}^m$. (That is, for each $x \in U$, the derivative $\varphi'(x)\colon (T_xM, \mathcal{O}_x) \to \mathbb{R}^m$ preserves orientation.)

An *oriented manifold* is a pair $(M, \mathcal{O})$, where M is a differentiable manifold and $\mathcal{O}$ is an orientation in M.

A manifold is said to be *orientable* when it is possible to define some orientation in it.

Let $\mathcal{O}$ be an orientation in a manifold M. We denote by $-\mathcal{O}$ the correspondence that associates to each $x \in M$ the orientation $-\mathcal{O}_x$ in T_xM, opposite to $\mathcal{O}_x$. It is easy to see that $-\mathcal{O}$ is an orientation of M, called *opposite orientation* of $\mathcal{O}$.

Example 8.1. The Euclidean space $\mathbb{R}^m$ is orientable. In fact, $\mathbb{R}^m$ is oriented: We always consider it with its natural orientation. More generally, every Lie group G is orientable: We choose an arbitrary orientation in the tangent space T_eG (e is the neutral element of G) and we extend it to each tangent space T_gG by requiring that the linear isomorphism $L_g \colon T_eG \to T_gG$, the derivative of the left translation $h \mapsto gh$, be positive at the point e. ◁

Example 8.2. Every open subset U of an orientable manifold M is an orientable manifold. In fact, for each $x \in U$, we have $T_xU = T_xM$. An orientation of M determines, in a natural way, an orientation in U, called the *induced orientation*. This example is a particular case of the next one. ◁

Example 8.3. Let $f \colon M \to N$ be a local diffeomorphism. An orientation $\mathcal{O}'$ in N determines, by means of f, an orientation $\mathcal{O}$ in M, characterized by the property of turning $f \colon (M, \mathcal{O}) \to (N, \mathcal{O}')$ into a positive map; that is, for each $x \in M$, the linear isomorphism $f'(x) \colon T_xM \to T_{f(x)}N$ preserves orientation. In fact, this condition defines the correspondence $x \mapsto \mathcal{O}_x$. In order to obtain a positive coordinate system $\varphi \colon U \to \mathbb{R}^m$ around the point $x \in M$, we just have to take a positive coordinate system $\psi \colon V \to \mathbb{R}^m$ around the point $y = f(x) \in N$, an open set $U \ni x$ in M that it is mapped diffeomorphically by f onto a subset of V, and we set $\varphi = \psi \circ f$. The orientation $\mathcal{O}$ is said to be *induced* by f. In particular, if N is orientable and there exists a local diffeomorphism $f \colon M \to N$, then M is orientable. ◁

Soon we will present examples of nonorientable manifolds.

Two coordinate systems, $\varphi \colon U \to \mathbb{R}^m$ and $\psi \colon V \to \mathbb{R}^m$, in a manifold M are said to be *compatible* when $U \cap V = \varnothing$, or when $U \cap V \neq \varnothing$ and the change of coordinates $\psi \circ \varphi^{-1} \colon \varphi(U \cap V) \to \psi(U \cap V)$ has a positive Jacobian determinant at every point of $\varphi(U \cap V)$. An atlas in M is said to be *coherent* when any two of its coordinate systems are compatible.

In an oriented manifold M, the set of positive coordinate systems is a coherent atlas. Since we are taking all of the positive systems, this is a *maximal* coherent atlas; that is, it is not a proper subset of any coherent atlas on M.

Conversely, if there exists a coherent atlas $\mathcal{A}$ in the manifold M, we define an orientation $\mathcal{O}_x$ in each tangent space T_xM by taking a coordinate system $\varphi \colon U \to \mathbb{R}^m$ that belongs to $\mathcal{A}$ and requiring that $\varphi'(x) \colon T_xM \to \mathbb{R}^m$ preserve orientation. The orientation of each tangent space T_xM is well defined because of the compatibility of the systems in $\mathcal{A}$. Note that the way that we defined $\mathcal{O}_x$ shows that we obtain an orientation of M.

Every coherent atlas in a manifold M is contained in a unique maximal coherent atlas. We could have defined, equivalently, an orientation of M as a maximal coherent atlas. The definition we gave has a better geometric flavor.

We say that an atlas $\mathcal{A}$, in an oriented manifold M, is *positive* when any coordinate systems belonging to $\mathcal{A}$ is positive relatively to the orientation of M. This means that $\mathcal{A}$ is contained in the maximal coherent atlas of M.

Proposition 8.1. *Let M, N be two oriented manifolds with the same dimension and $f\colon M \to N$ be a local diffeomorphism. The set of points $x \in M$ at which the derivative $f'(x)\colon T_xM \to T_{f(x)}N$ is positive is an open subset of M.*

Proof. Given $x \in M$ and $y = f(x) \in N$, let $\varphi\colon U \to \mathbb{R}^m$ and $\psi\colon V \to \mathbb{R}^m$ be coordinate systems in M and N respectively, with $x \in U$, $y \in V$ and $f(U) \subset V$. Then $f'(x)\colon T_xM \to T_yN$ is positive if, and only if, $(\psi \circ f \circ \varphi^{-1})'(\varphi(x))\colon \mathbb{R}^m \to \mathbb{R}^m$ has a positive Jacobian. Since this Jacobian is a continuous function of x, this concludes the proof of the proposition. □

Corollary 8.1. *Let M and N be oriented manifolds. If M is connected, then a local diffeomorphism $f\colon M \to N$ is either positive or negative.*

In fact, the set of all points $x \in M$ at which the derivative $f'(x)\colon T_xM \to T_yN$ reverses the orientation is also open. Since this set and the set of Proposition 8.1 are disjoint, one of them is empty by virtue of the connectedness of M.

Corollary 8.2. *Let $\varphi\colon U \to \mathbb{R}^m$ be a coordinate system in an oriented manifold M. If the domain U is connected, then φ is either positive or negative.*

Corollary 8.3. *In a connected oriented manifold, there are two possible orientations.*

In fact, consider the orientations $\mathcal{O}$ and $\mathcal{O}'$ in a connected manifold M. The identity map $f\colon (M, \mathcal{O}) \to (M, \mathcal{O}')$ is a local diffeomorphism. Hence, either f is positive (and in this case, $\mathcal{O} = \mathcal{O}'$) or f is negative (and then $\mathcal{O}' = -\mathcal{O}$).

Corollary 8.4. *Suppose that, in a manifold M, there exist coordinate systems $\varphi\colon U \to \mathbb{R}^m$, $\psi\colon V \to \mathbb{R}^m$, with connected domains U, V, such that at two points of $\varphi(U \cap V)$ the change of coordinates $\psi \circ \varphi^{-1}\colon \varphi(U \cap V) \to \psi(U \cap V)$ has Jacobian determinants with distinct signs. Then M is nonorientable.*

Remark. In the case of Corollary 8.4, the intersection $U \cap V$ is necessarily disconnected.

Example 8.4. If M and N are orientable manifolds, the same happens with their product $M \times N$. In fact, let $\mathcal{A}$ and $\mathcal{B}$ be the coherent atlas, respectively, in M and N, defining the orientations of these manifolds. The atlas $\mathcal{A} \times \mathcal{B}$ is coherent because if $\varphi_1, \varphi_2 \in \mathcal{A}$ e $\psi_1, \psi_2 \in \mathcal{B}$, then $(\varphi_2 \times \psi_2) \circ (\varphi_1 \times \psi_1)^{-1} = (\varphi_2 \circ \varphi_1^{-1}) \times (\psi_2 \circ \psi_1^{-1})$ has as Jacobian determinant the product of the (positive) Jacobians of $\varphi_2 \circ \varphi_1^{-1}$ and $\psi_2 \circ \psi_1^{-1}$.

The orientation defined in $M \times N$ by the atlas $\mathcal{A} \times \mathcal{B}$ is called *product* of the orientations of M and N (in this order). If $(u_1, \ldots, u_m)$ and $(v_1, \ldots, v_m)$ are positive bases in T_xM and T_yN respectively, then $(u_1, \ldots, u_m, v_1, \ldots, v_m)$ is a positive basis in $T_{(x,y)}(M \times N)$.

Conversely, if the product $M \times N$ is an orientable manifold, then each of the manifolds M, N is orientable. In fact, fix an orientation in $M \times N$ and a coordinate system $\overline{\psi}\colon V \to \mathbb{R}^n$ in N, whose domain V is connected. For each $x \in M$, take a coordinate system $\varphi\colon U \to \mathbb{R}^m$ in M, with $x \in U$, such that $\varphi \times \overline{\psi}$ is positive in $M \times N$. The systems thus obtained form an atlas $\mathcal{A}$ in M. We claim that $\mathcal{A}$ is coherent. In fact, if $\varphi_1, \varphi_2 \in \mathcal{U}$ then the Jacobian of

$$(\varphi_2 \times \overline{\psi}) \circ (\varphi_1 \times \overline{\psi})^{-1} = (\varphi_2 \circ \varphi_1^{-1}) \times (\overline{\psi} \circ \overline{\psi}^{-1}) = (\varphi_2 \circ \varphi_1^{-1}) \times id$$

is positive at each point $(\varphi_1(x), \overline{\psi}(y))$. But this equals the Jacobian of $\varphi_2 \circ \varphi_1^{-1}$ at the point $\varphi_1(x)$. In a similar way, we prove that N is orientable. ◁

Consider a fixed point $b \in N$ and a basis $(v_1, \ldots, v_n)$ in T_bN. For each $x \in M$, a basis $(u_1, \ldots, u_m)$ in T_xM is positive if, and only if, the basis $(u_1, \ldots, u_m, v_1, \ldots, v_m)$ is positive in $T_{(x,b)}(M \times N)$.

Example 8.5. The map $f\colon S^m \times \mathbb{R} \to \mathbb{R}^{m+1} - \{0\}$, defined by $f(x,t) = e^t \cdot x$, is a diffeomorphism of the product $S^m \times \mathbb{R}$ onto the open subset $\mathbb{R}^{m+1} - \{0\}$ of the Euclidean space. Hence, $S^m \times \mathbb{R}$ is orientable. It follows from the previous example that the sphere S^m is orientable. Given $x \in S^m$, a basis $(v_1, \ldots, v_m)$ of T_xS^m is positive if, and only if $\det[x, v_1, \ldots, v_m] > 0$. ◁

Example 8.6. Let $A = (0,5) \times (0,1)$ be the open rectangle with base 5 and height 1. Given two integers $i < j$ in the interval $[0,5]$, we let $A_{ij} = (i,j) \times (0,1)$ be a rectangle with base $j - i$ and height 1.

The *Moebius band* M is the quotient space of A by the equivalence relation that identifies each point $(s,t) \in A_{01}$ with $(s+4, 1-t) \in A_{45}$ (see Figure 8.1). Let $\pi\colon A \to M$ be the quotient map. The restrictions $\pi|A_{03}$ e $\pi|A_{25}$ are, respectively, homeomorphisms onto open sets U and V in M. We denote by $\varphi\colon U \to A_{03}$ and $\psi\colon V \to A_{25}$ their inverses. We see that

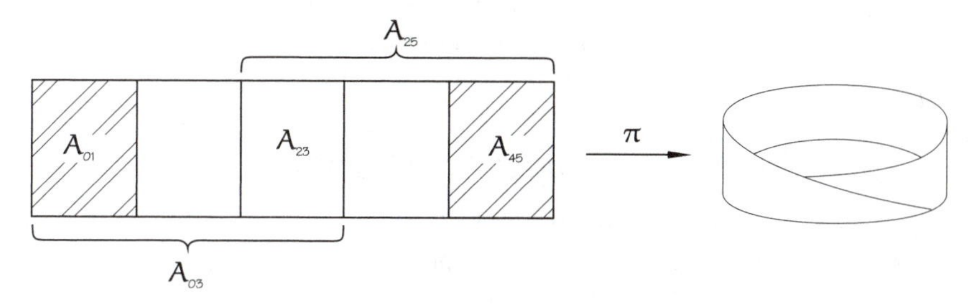

Figure 8.1. The Moebius band.

$\varphi(U \cap V) = A_{01} \cup A_{23}$ and $\psi(U \cap V) = A_{23} \cup A_{45}$. Moreover, the change of coordinates

$$\psi \circ \varphi^{-1} \colon A_{01} \cup A_{23} \to A_{23} \cup A_{45}$$

is the identity in A_{23} and it is given by $(s, t) \mapsto (s+4, 1-t)$ in A_{01}. It follows from Corollary 8.4 that the Moebius band is a nonorientable manifold. ◁

Example 8.7. The antipodal map $\alpha \colon S^m \to S^m$, $\alpha(x) = -x$ is a diffeomorphism, with $\alpha^{-1} = \alpha$. Let us check whether α preserves or reverses the orientation of S^m. Given $x \in S^m$, we set $E_x = T_x S^m$. We have $E_x = E_{-x}$. A basis $(v_1, \ldots, v_m)$ em E_x is positive if, and only if, $\det[x, v_1, \ldots, v_m] > 0$. It results from this that, although the nonoriented vector spaces E_x and E_{-x} are the same, we have $\mathcal{O}_{-x} = -\mathcal{O}_x$; that is, the orientations of E_x and E_{-x} do not coincide. The derivative $\alpha'(x) \colon E_x \to E_{-x}$ is given by the multiplication by -1. With respect to the orientations $\mathcal{O}_x$ and $\mathcal{O}_{-x}$ adopted in these spaces, $\alpha'(x)$ is a positive isomorphism if, and only if, m is odd. Thus, the antipodal map $\alpha \colon S^m \to S^m$ preserves orientation when m is odd and reverses it when m is even. ◁

Example 8.8. We prove now that the real projective space P^m is orientable when m is odd and that it is nonorientable when m is even. With this in mind, consider the canonical projection $\pi \colon S^m \to P^m$, which is a local diffeomorphism, and the antipodal map $\alpha \colon S^m \to S^m$. We have $\pi \circ \alpha = \pi$. If m is odd, we define an orientation in each tangent space $T_y P^m$, $y = \pi(x)$, by requiring that the linear isomorphism $\pi'(x) \colon T_x S^m \to T_y P^m$ be positive. It seems that there is an ambiguity, because we also have $y = \pi(-x)$. But, since $\pi'(-x) \circ \alpha'(x) = \pi'(x)$ and $\alpha'(x)$ is positive, the isomorphism $\pi'(-x)$ would induce the same orientation in $T_y P^m$. This defines an orientation in P^m.

Conversely, assume that P^m is orientable. Since S^m is connected, we can (see Corollary 8.1) choose the orientation of P^m in such a way that

$\pi\colon S^m \to P^m$ is positive. Now we fix $x \in S^m$. Since the isomorphisms $\pi'(-x)$ and $\pi'(x)$ are both positive, it follows that $\alpha'(x) = \pi'(-x)^{-1} \circ \pi'(x)$ is positive and therefore m is odd, according to the previous example. ◁

8.3 Properly Discontinuous Groups of Diffeomorphisms

Let $f\colon M \to N$ be a local diffeomorphism. When N is orientable, we know that f induces an orientation in M. Consider now the inverse situation: Supposing that M is orientable, is it possible to define, by using f, an orientation in N? The particular case $\pi\colon S^m \to P^m$ was solved in Example 8.8. The hypothesis that f is surjective is, evidently, necessary.

Proposition 8.2. *Let $f\colon M \to N$ be a surjective local diffeomorphism, defined on a connected oriented manifold. In order that N be orientable, it is necessary and sufficient that, for any x, $y \in M$ with $f(x) = f(y)$, the linear isomorphism $f'(y)^{-1} \circ f'(x)\colon T_xM \to T_yM$ be positive.*

Proof. If the condition holds, we define an orientation in N by taking, for each $b \in N$, a point $x \in f^{-1}(b)$ and imposing that the linear isomorphism $f'(x)\colon T_xM \to T_bN$ be positive. The admitted condition means that the orientation thus defined in each T_bN does not depend on the choice of the point x in $f^{-1}(b)$. Moreover, if we take in M a positive coordinate system $\varphi\colon U \to \mathbb{R}^m$, defined in an open set $U \ni x$ which is mapped diffeomorphically onto an open set $V \ni b$, the composite map $\psi = \varphi \circ f^{-1}\colon V \to \mathbb{R}^m$ is a positive coordinate system in N. This shows that we have indeed an orientation in N.

Conversely, let M, N be oriented manifolds. (Here we use the connectedness of M.) Then $f\colon M \to N$ is either positive or negative. By changing, if necessary, the orientation of N, we may suppose that f is positive. Then, for any x, $y \in M$ with $f(x) = f(y)$, the linear isomorphisms $f'(x)\colon T_xM \to T_{f(x)}N$, $f'(y) = T_yM \to T_{f(y)}N$ are positive and consequently the isomorphism $f'(y)^{-1} \circ f'(x)$ is also positive. □

A frequent situation where Proposition 8.2 applies is that of a properly discontinuous group of diffeomorphisms, as we explain now.

Proposition 8.3. *Let M be a connected manifold of class C^k and G be a properly discontinuous group of diffeomorphisms of class C^k in M. If the quotient space M/G is Hausdorff then there exists a unique manifold structure of class C^k in M/G such that the quotient map $\pi\colon M \to M/G$ is a local diffeomorphism of class C^k. Suppose that M is oriented. In order that M/G be orientable, it is necessary and sufficient that each diffeomorphism belonging to G preserve orientation.*

Proof. Consider an open covering of M such that each open set contains at most one point of each orbit of G. Let U, V be two of these open sets. Then $\pi|U$ and $\pi|V$ are homeomorphisms onto open sets $U_0, V_0 \subset M/G$. Suppose that $U_0 \cap V_0 \neq \varnothing$. Let $A = (\pi|U)^{-1}(U_0 \cap V_0)$ and $B = (\pi|V)^{-1}(U_0 \cap V_0)$. We claim that the homeomorphism $\xi = (\pi|V)^{-1} \circ (\pi|U) \colon A \to B$ is of class C^k. In order to prove this, we remark that if $x \in A$, then $\xi(x) = y \Rightarrow \pi(x) = \pi(y) \Rightarrow y = \alpha(x)$ for some $\alpha \in G$. That is, for all $x \in A$, there exists $\alpha \in G$ such that $\xi(x) = \alpha(x)$. By fixing $x \in A$, let Z be an open set such that $x \in Z \subset A$ and $\xi(Z) \subset \alpha(A)$. We know that $\beta \neq \alpha \Rightarrow \beta(A) \cap \alpha(A) = \varnothing$. Thus, $\xi(Z) \cap \beta(Z) = \varnothing$ for all $\beta \neq \alpha$. This shows that $\xi(y) \neq \beta(y)$ for all $y \in Z$ and therefore, $\xi|Z = \alpha|Z$. It follows that $\xi \in C^k$ in Z. Since Z is a neighborhood of an arbitrary point $x \in A$, we conclude that $\xi \colon A \to B$ is of class C^k. Thus, the topological space M/G is covered by open domains of homeomorphisms $\varphi = (\pi|U)^{-1} \colon U_0 \to U$, which take values in open subsets of M, in such a way that, when the domain of $\psi \colon V_0 \to V$ intersects that of φ, then the change of coordinates $\xi = \psi \circ \varphi^{-1} \colon \varphi(U_0 \cap V_0) \to \psi(U_0 \cap V_0)$ is of class C^k. Since φ is open, M/G inherits from M a countable basis. Thus the homeomorphisms φ define a manifold structure of class C^k in M/G. Evidently, the quotient map $\pi \colon M \to M/G$ is a local diffeomorphism. The uniqueness of the structure in M/G follows easily from this property. Suppose now that M is oriented and each $\alpha \in G$ is a positive diffeomorphism of M. Then the local diffeomorphism $\pi \colon M \to M/G$ satisfies $\pi(x) = \pi(y) \Rightarrow y = \alpha(x)$, with $\alpha \in G$. Since $\pi \circ \alpha = \pi$, we conclude that $\pi'(y) \circ \alpha'(x) = \pi'(x)$; that is, $\pi'(y)^{-1} \circ \pi'(x) = \alpha'(x)$, which is a positive linear isomorphism. It follows from Proposition 8.2 that M/G is orientable.

Conversely, if M/G is orientable, we take arbitrarily $\alpha \in G$ and $x \in M$. Let $y = \alpha(x)$. Then $\pi(x) = \alpha(y)$. By Proposition 8.2, the isomorphism $\pi'(y)^{-1} \circ \pi'(x)$ is positive. But this isomorphism coincides with $\alpha'(x)$. It follows that α is positive, which completes the proof. □

Example 8.9. Now we will see that the orientability of the projective space (P^m is orientable if, and only if, m is odd) is explained more generally by Proposition 8.3. In Example 7.8, where we have the group of translations of $\mathbb{R}^n$ by vectors with integral coordinates, the quotient space is the n-dimensional torus $\mathbb{R}^n/\mathbb{Z}^n = S^1 \times \ldots \times S^1$ (n factors). Since each translation $x \mapsto x + v$ is a positive diffeomorphism of $\mathbb{R}^n$, we conclude that the n-dimensional torus is orientable, a fact that we already knew, because it is the product of n orientable manifolds, or because it is a Lie group. ◁

Example 8.10. Let $M = S^1 \times \mathbb{R}$. The diffeomorphism $h \colon M \to M$, defined by $h(x, y, z) = (x, -y, z + 1)$, generates a cyclic group $G = \{h^n; n \in \mathbb{Z}\}$ of

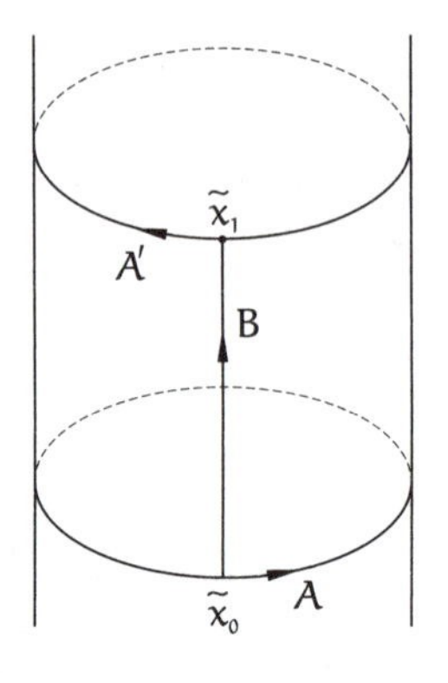

Figure 8.2.

diffeomorphisms of M, which is properly discontinuous. We can imagine M as a vertical cylinder in $\mathbb{R}^3$. The diffeomorphism h maps each horizontal circle of M onto the circle located one unity above it but, when doing this, it also reflects the circle around a diameter. The n-th iterate h^n is a positive diffeomorphism of M if n is even and it is negative when n is odd. It follows that the quotient space M/G is a nonorientable manifold. It is called the *Klein bottle.* ◁

Example 8.11. In the example above, let $p\colon M \to M/G$ be the canonical covering map of the cylinder M onto the Klein bottle M/G. The fundamental group of the cylinder is cyclic infinite, generated by the homotopy class of the path $A(s) = (\cos 2\pi s, \sin 2\pi s, 0)$. Consider now the paths $A'(s) = (\cos 2\pi s, -\sin 2\pi s, 1)$, $B(s) = (1, 0, s)$, in the cylinder, and their images $a = p \circ A = p \circ A'$ and $b = p \circ B$ in the Klein bottle (see Figure 8.2). Both a and b are closed paths, with base at the point $x_0 = p(\tilde{x}_0) = p(\tilde{x}_1)$, where $\tilde{x}_0 = (1, 0, 0)$ and $\tilde{x}_1 = (1, 0, 1)$. We claim that we do not have $ab \cong ba$ in the Klein bottle. In fact, by taking liftings starting at the point $\tilde{x}_0$, we have $\widetilde{ab} = AB \simeq A$ and $\widetilde{ba} = BA' \simeq A' \simeq A^{-1}$ (free homotopies). Since A and A^{-1} are not freely homotopic in the cylinder, it follows that $\widetilde{ab}$ is not homotopic to $\widetilde{ba}$ in this cylinder. Consequently, we do not have $ab \cong ba$ in the Klein bottle. (See Proposition 6.10.) Thus, we have proved, once more, that the fundamental group of the Klein bottle is not commutative. ◁

8.4 Oriented Double Covering

An *oriented double covering* is a map $p\colon \widetilde{M} \to M$, of class C^k, with the following properties:

1. M is a connected manifold, $\widetilde{M}$ is an oriented manifold, and p is a local diffeomorphism;

2. For each $y \in M$, the inverse image $p^{-1}(y)$ contains exactly two points;

3. If $p(x_1) = p(x_2)$ with $x_1 \neq x_2$, then the linear isomorphism $p'(x_2)^{-1} : p'(x_1) : T_{x_1}\widetilde{M} \to T_{x_2}\widetilde{M}$ is negative.

By virtue of Proposition 6.5, an oriented double covering $p\colon \widetilde{M} \to M$ is a proper covering map.

Sometimes we say, rather incorrectly, that $\widetilde{M}$ (not p) is an oriented double covering of M.

Example 8.12. When m is even, the quotient map $\pi\colon S^m \to P^m$ is an oriented double covering of the projective space P^m. When m is odd, π does not satisfy the Condition 3. above. ◁

Example 8.13. Let $\alpha\colon \widetilde{M} \to \widetilde{M}$ be a negative involution ($\alpha \circ \alpha = id$), of class C^k, without fixed points, on a connected oriented manifold. Then $\{\alpha, id\}$ is a properly discontinuous group of diffeomorphisms of $\widetilde{M}$. We indicate by $\widetilde{M}/\alpha$ the quotient manifold. (See Proposition 8.3.) The quotient map $\pi\colon \widetilde{M} \to \widetilde{M}/\alpha$ is an oriented double covering. An example of this situation is $\widetilde{M} = S^1 \times S^1 =$ the two-dimensional torus. We define $\alpha\colon \widetilde{M} \to \widetilde{M}$ by setting $\alpha(z, w) = (\bar{z}, -w)$, where $\bar{z}$ is the complex conjugate of z. Then α is a negative involution without fixed points in the torus $\widetilde{M}$. The quotient manifold $\widetilde{M}/\alpha$ is diffeomorphic to the Klein bottle. The canonical projection $\pi\colon \widetilde{M} \to \widetilde{M}/\alpha$ shows that the torus is an oriented double covering of the Klein bottle. Later, we will show that every oriented double covering is essentially obtained in this way. ◁

Example 8.14. (Product covering) Let M be a connected oriented manifold. Take $M_1 = M \times \{1\}$ and $M_2 = M \times \{2\}$. Then $\widetilde{M} = M_1 \cup M_2$ is a manifold, the disjoint union of two diffeomorphic copies of M. Now define $p\colon \widetilde{M} \to M$ by setting $p(x, 1) = p(x, 2) = x$. Let's choose an orientation of $\widetilde{M}$ by requiring that $p|M_1$ be positive and $p|M_2$ negative. Then $p\colon \widetilde{M} \to M$ is an oriented double covering, called the *product covering.* ◁

More generally, we say that an oriented double covering $p\colon \widetilde{M} \to M$ is *trivial* when $\widetilde{M} = \widetilde{M}_1 \cup \widetilde{M}_2$ is the disjoint union of two open subsets, such that each one of them is mapped by p diffeomorphically onto M. We now show that this is essentially the only possible oriented double covering when the base space M is orientable.

It is important to note that if $\widetilde{M}$ is connected, then the base M of the oriented double covering $p\colon \widetilde{M} \to M$ must be a nonorientable manifold, according to Proposition 8.2. Observe also that $p\colon S^1 \to S^1$, $p(z) = z^2$ is a covering with two leaves, but it is not an oriented double covering.

Proposition 8.4. *Let $p\colon \widetilde{M} \to M$ be an oriented double covering. If $U \subset M$ is an oriented open set, then $p^{-1}(U) = \widetilde{U}_1 \cup \widetilde{U}_2$, is the disjoint union of two open sets, such that p maps each one of them diffeomorphically onto U. In $\widetilde{U}_1$, p is positive and in $\widetilde{U}_2$, it is negative.*

Proof. Let $\widetilde{U}_1 = \{x \in p^{-1}(U); p'(x) < 0\}$ and $\widetilde{U}_2 = \{x \in p^{-1}(U); p'(x) > 0\}$. Evidently, $p^{-1}(U) = \widetilde{U}_1 \cup \widetilde{U}_2$, the union of two disjoint open sets. In each $\widetilde{U}_i$, p is injective because $p(x_1) = p(x_2)$ with $x_1 \neq x_2$ would force $p'(x_1)$ and $p'(x_2)$ to have opposite signs. Moreover, each point $y \in U$ is the image of two points $x_1, x_2 \in p^{-1}(U)$; in one of them, the derivative of p is positive and in the other, it is negative. Hence, $p(\widetilde{U}_1) = p(\widetilde{U}_2) = U$. We conclude that $p|\widetilde{U}_1$ and $p|\widetilde{U}_2$ are bijections (and therefore, diffeomorphisms) onto U. □

Proposition 8.5. *Let $p\colon \widetilde{M} \to M$ be an oriented double covering. The following statements are equivalent:*

1. *M is orientable;*
2. *$\widetilde{M}$ is disconnected;*
3. *The covering $p\colon \widetilde{M} \to M$ is trivial.*

Proof. $1 \Rightarrow 2$: This follows from Proposition 8.2.

$2 \Rightarrow 3$: Suppose that $\widetilde{M}$ is disconnected and take a connected component C of $\widetilde{M}$. Since p is a proper local diffeomorphism, the image $p(C)$ of the open-closed set C is open and closed in the connected manifold M. Hence, $p(C) = M$. Thus, p maps each connected component of $\widetilde{M}$ onto M. Since the inverse image by p of each point of M has two points, we conclude that $\widetilde{M}$ cannot have more than two components. As a consequence, $\widetilde{M} = \widetilde{M}_1 \cup \widetilde{M}_2$ has precisely two connected components and p is injective in each one of them. Thus, $p|M_1$ and $p|M_2$ are diffeomorphisms onto M; that is, p is trivial.

$3 \Rightarrow 1$: Obvious. □

Corollary 8.5. *Let $p\colon \widetilde{M} \to M$ be an oriented double covering. $\widetilde{M}$ is connected if, and only if, M is nonorientable.*

Now we prove the uniqueness of the oriented double covering.

Proposition 8.6. *Let $p_1\colon \widetilde{M}_1 \to M$ and $p_2\colon \widetilde{M}_2 \to M$ be oriented double coverings of the same manifold M. There exists a unique positive diffeomorphism $f\colon \widetilde{M}_1 \to \widetilde{M}_2$ such that $p_2 \circ f = p_1$. This is illustrated by the commutative diagram below.*

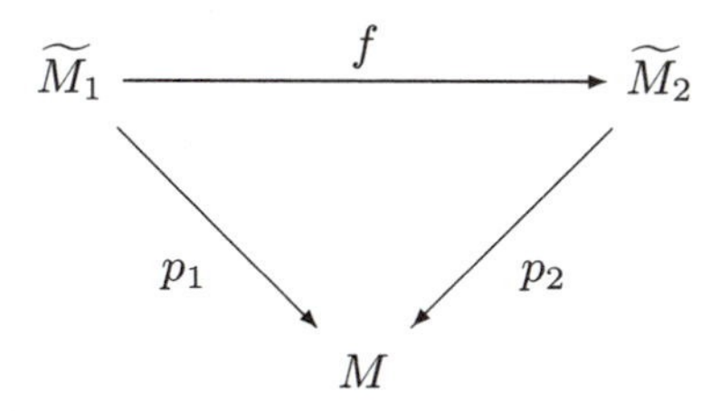

Proof. The conditions that f be positive and satisfy $p_2 \circ f = p_1$ already define a bijection $f\colon \widetilde{M}_1 \to \widetilde{M}_2$: For each point $x \in \widetilde{M}_1$, $f(x) = y$ is the point of $\widetilde{M}_2$, which is mapped by p_2 onto the point $z = p_1(x)$ in such a way that $p_2'(y)^{-1} \circ p_1'(x)$ becomes a positive linear isomorphism. (There exist two points of M_2 which are mapped by p_2 onto the point z, but only one of them satisfies the last condition.) Now we just have to prove that $f \in C^k$ if p_1 and p_2 are of class C^k. We fix $x \in \widetilde{M}_1$. Let $U \ni z = p_1(x)$ be a domain of a coordinate system in M; we orient U in such a way that $p_1^{-1}(U) = \widetilde{U}_1 \cup \widetilde{V}_1$ with $x \in \widetilde{U}_1$ and $p_1|\widetilde{U}_1$ be a positive diffeomorphism onto U. Then $p_2^{-1}(U) = \widetilde{U}_2 \cup \widetilde{V}_2$, where $p_2|\widetilde{U}_2$ is a positive diffeomorphism onto U. It follows that $y = f(x) \in \widetilde{U}_2$ and $f|\widetilde{U}_1 = (p_2|\widetilde{U}_2)^{-1} \circ (p_1|\widetilde{U}_1)$. □

The corollary below shows that every oriented double covering is essentially obtained by taking the quotient space of a manifold by a negative involution without fixed points.

Corollary 8.6. *Let $p\colon \widetilde{M} \to M$ be an oriented double covering. There exists a unique negative involution $\alpha\colon \widetilde{M} \to \widetilde{M}$, of class C^k, such that $p \circ \alpha = p$. The involution α does not have fixed points. There exists a unique diffeomorphism $\xi\colon \widetilde{M}/\alpha \to M$ such that $p = \xi \circ \pi$, where $\pi\colon \widetilde{M} \to \widetilde{M}/\alpha$ is the quotient map.*

In fact, let $\widetilde{M}_1$ and $\widetilde{M}_2$ be the same manifold $\widetilde{M}$ with two opposite orientations. The map p determines two oriented double coverings $p_i\colon \widetilde{M}_i \to M$ $(i = 1, 2)$, that differ from the original one only by the orientation of their domains. By Proposition 8.6, there exists a unique positive diffeomorphism $\alpha\colon \widetilde{M}_1 \to \widetilde{M}_2$ such that $p_2 \circ \alpha = p_1$. Going back to the original notation, $\alpha\colon \widetilde{M} \to \widetilde{M}$ is a diffeomorphism that reverses the orientation of $\widetilde{M}$; this, along with the equality $p \circ \alpha = p$, shows that, for each $x \in \widetilde{M}$, $\alpha(x) = y$ is the other point of $\widetilde{M}$ such that $p(x) = p(y)$. Thus, α does not have fixed points and $\alpha \circ \alpha = id$; that is, α is an involution. The fact that $\alpha \in C^k$ was proved in Proposition 8.6. Finally, considering the quotient space $\widetilde{M}/\alpha$, since the equivalence relation defined by p in $\widetilde{M}$ is the same defined by $\pi\colon \widetilde{M} \to \widetilde{M}/\alpha$, it follows that there exists a continuous bijection $\xi\colon \widetilde{M}/\alpha \to M$ such that $\xi \circ \pi = p$, as illustrated by the commutative diagram below.

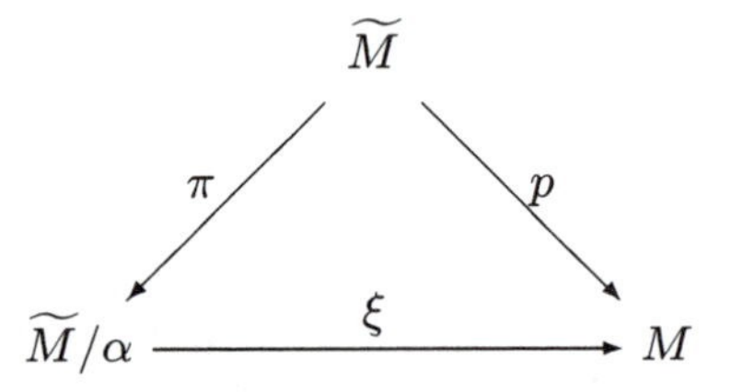

Since p is a local diffeomorphism of class C^k, it follows that ξ is a diffeomorphism of class C^k.

Proposition 8.7. *Every connected manifold M of class C^k has an oriented double covering.*

Proof. Let $\widetilde{M}$ be the set of ordered pairs $(x, \mathcal{O}_x)$ where $x \in M$ and $\mathcal{O}_x$ is an orientation in the tangent space T_xM. We define a map $p\colon \widetilde{M} \to M$ by setting $p(x, \mathcal{O}_x) = x$. Clearly, for each $x \in M$, the inverse image $p^{-1}(x)$ contains exactly two points: $(x, \mathcal{O}_x)$ and $(x, -\mathcal{O}_x)$. We introduce now a manifold structure of class C^k in $\widetilde{M}$ in such a way that p becomes an oriented double covering of class C^k. For each oriented open set $U \subset M$, let $\widetilde{U}$ be the set of pairs $(x, \mathcal{O}_x)$ such that $x \in U$, and $\mathcal{O}_x$ is the orientation of U at the point x. The map $\varphi_U : \widetilde{U} \to U$, defined by $\varphi_U = p|\widetilde{U}$, is a bijection. The domains $\widetilde{U}$ of these bijections φ_U cover the set $\widetilde{M}$. Given $\varphi_U : \widetilde{U} \to U$ and $\varphi_V\colon \widetilde{V} \to V$, if $\widetilde{U} \cap \widetilde{V} \neq \varnothing$ then $\varphi_U|(\widetilde{U} \cap \widetilde{V}) = \varphi_V|(\widetilde{U} \cap \widetilde{V})$; hence the "change of coordinates" $\varphi_V \circ \varphi_U^{-1} : \varphi_U(\widetilde{U} \cap \widetilde{V}) \to \varphi_V(\widetilde{U} \cap \widetilde{V})$ is simply the identity map. Therefore, the atlas constituted by the bijections φ_U determines in $\widetilde{M}$ a manifold structure of class C^k, and $p\colon \widetilde{M} \to M$ is a local diffeomorphism with respect to this structure. Such structure is defined by the condition that each φ_U be a diffeomorphism. In order to show that $\widetilde{M}$ is orientable, we remark that $\widetilde{M}$ has a natural orientation, imposed by its definition: in each point $\tilde{x} = (x, \mathcal{O}_x) \in \widetilde{M}$, consider the orientation $\mathcal{O}_{\tilde{x}}$ which makes the linear isomorphism $p'(\tilde{x})\colon (T_{\tilde{x}}\widetilde{M}, \mathcal{O}_{\tilde{x}}) \to (T_xM, \mathcal{O}_x)$ positive. The map $p\colon \widetilde{M} \to M$ is an oriented double covering. $\square$

Corollary 8.7. *Every simply connected manifold is orientable.*

In fact, let M be a simply connected manifold. Every covering of M with connected domain is a homeomorphism. Hence, the oriented double covering of M is disconnected; therefore, M is orientable.

8.5 Relations with the Fundamental Group

Let $p\colon \widetilde{M} \to M$ be an oriented double covering. Given a path $a\colon I \to M$ and a point $\tilde{x} \in \widetilde{M}$ such that $p(\tilde{x}) = a(0)$, there exists a unique path $\tilde{a}\colon I \to \widetilde{M}$ such that $p \circ \tilde{a} = a$ and $\tilde{a}(0) = \tilde{x}$. If we adopt for $\widetilde{M}$ the model presented in the proof of Proposition 8.7, we have $\tilde{x} = (x, \mathcal{O}_x)$ and the path $\tilde{a}$ can be interpreted as the continuation of the orientation $\mathcal{O}_x$, by continuity, along the path a. In fact, we have $\tilde{a}(s) = (a(s), \mathcal{O}_{a(s)})$ for all $s \in I$. Since $\tilde{a}(s)$ depends continuously on s, it is natural to say that the orientation $\mathcal{O}_{x(s)}$ also depends continuously on the parameter s.

Let $b\colon I \to M$ be another path in M with the same endpoints as a. If $a \cong b$ (that is, if a and b are homotopic with the endpoints fixed in M), then their liftings $\tilde{a}$ and $\tilde{b}$ starting from the same point $\tilde{x}$ are also homotopic with fixed endpoints in $\widetilde{M}$. In particular, $\tilde{a}(1) = \tilde{b}(1)$. This means that, starting from an orientation $\mathcal{O}_x$ in T_xM and extending it by continuity along the two homotopic paths with fixed endpoints, we obtain in the final the same orientation. In particular, if the manifold M is simply connected, it is possible to orient it by choosing an orientation $\mathcal{O}_{x_0}$ at a fixed point $x_0 \in M$ and, given any point $x \in M$, we connect x to x_0 using a path in M and we extend $\mathcal{O}_{x_0}$ continuously along this path. The orientation $\mathcal{O}_x$ thus obtained does not depend on the path chosen in order to connect x_0 to x because, since M is simply connected, any two of these paths are homotopic with fixed endpoints.

Consider now closed paths in M, with base at a point x_0. Given the oriented double covering $p\colon \widetilde{M} \to M$, the lifting of a closed path may be closed or open. In terms of the model of Proposition 8.7: extending continuously an orientation $\mathcal{O}_{x_0}$ along the path a, with $a(0) = a(1) = x_0$, it is possible to obtain, in the final, the orientation $\mathcal{O}_{x_0}$ or the opposite orientation $-\mathcal{O}_{x_0}$. This fact depends only on the path a, but not on the orientation $\mathcal{O}_{x_0}$. In the first case, we say that a is an *orienting path.* In the second case (in which, by extending $\mathcal{O}_{x_0}$ along a, we obtain in the end the opposite orientation $-\mathcal{O}_{x_0}$) we say that a is a *disorienting path.*

A manifold M is orientable if, and only if, every closed path in M is an orienting path. In the projective plane P^2, a projective line (image of half a great circle by the projection $\pi\colon S^2 \to P^2$) is a disorienting path.

Every closed path homotopic to a constant is an orienting path. In particular, if $\varphi\colon U \to \mathbb{R}^m$ is a coordinate system in M, and $\varphi(U)$ is the Euclidean ball, then every closed path contained in U is an orienting path. (We say then: Every sufficiently small closed path is an orienting path.)

The central circle of a Möbius ban is a disorienting path.

An interesting conjecture in cosmology states that the universe is an orientable manifold. Otherwise, a person who took a trip along a disorienting path would return mirrored: with the heart on the right side, writing everything in the opposite order and with the other hand. The arrows of his clock would move in the opposite sense and any books he carried with him would be illegible for us. On the other hand, he would think that everything here had changed while he was travelling.

Consider a nonorientable connected manifold M. Let a, b be two closed paths in M, with base in x_0. If $a \cong b$, then a is an orienting path if, and only if, b is also an orienting path. If two closed paths are orienting paths, their product is also an orienting path and so are their inverses. Thus, the homotopy classes of the orienting paths constitute a subgroup $H \subset \pi_1(M, x_0)$. H is the image of the fundamental group of $\widetilde{M}$ by the induced homomorphism $p_{\#}\colon \pi_1(\widetilde{M}, \tilde{x}_0) \to \pi_1(M, x_0)$. (It does not matter which is the chosen point $\tilde{x}_0$ over x_0, because every covering with two leaves is regular.)

In particular, we conclude that the fundamental group of a nonorientable manifold always has a subgroup of index 2.

8.6 Exercises

1. Let $\alpha\colon S^1 \times \mathbb{R} \to S^1 \times \mathbb{R}$ be defined by $\alpha(z,t) = (-z, -t)$. Show that α is a negative diffeomorphism, with $\alpha \circ \alpha = id$, that $G = \{id, \alpha\}$ is a properly discontinuous group and that $M = (S^1 \times \mathbb{R})/G$ is the Möbius band. Conclude again that M is nonorientable.

2. In a differentiable manifold M, the domain of a coordinate system is an orientable submanifold, even when M is nonorientable.

3. Every complex analytic manifold is orientable.

4. If the fundamental group of a connected manifold has seven or nine elements, the manifold is orientable.

5. If, for each point of a spherical surface, there exists a straight line that varies continuously with the point, prove that at least one of these lines passes through the center of the sphere.

In the following exercises we use the notation below:

- $M \subset \mathbb{R}^m$ denotes a surface of dimension m and class C^∞.
- $TM = \{(x, v) \in \mathbb{R}^m \times \mathbb{R}^n; x \in M, v \in T_x M\}$ = *tangent bundle* of M.
- $T^1 M = \{(x, u) \in TM; |u| = 1\}$ = *unit tangent bundle* of M.
- $\Delta M = \{(x, [u]); (x, u) \in T^1 M, [u] = \{u, -u\}\}$ = *tangent direction bundle* of M.

6. Prove that TM and $T^1 M$ are surfaces of class C^∞ in $\mathbb{R}^n \times \mathbb{R}^n$, with dimensions $2m$ and $2m - 1$, respectively, both of them orientable.

7. Prove that the maps $\pi: TM \to M$ and $\pi^1 = \pi|T^1 M: T^1 M \to M$ are locally trivial fibrations with typical fiber $\mathbb{R}^m$ and S^{m-1} respectively.

8. Prove that the fibrations $\pi^1: T^1 S^2 \to S^2$ and $\pi: SO(3) \to S^2$ (this last one was considered in Chapter 3) are equivalent; that is, there exists a diffeomorphism $\varphi: T^1 S^2 \to SO(3)$ such that $\pi \circ \varphi = \pi^1$. In particular, the fundamental group of $T^1 S^2$ is $\mathbb{Z}_2$.

9. Prove that ΔM is the quotient space of $T^1 M$ by the involution $(x, u) \mapsto (x, -u)$; hence, it is a differentiable manifold and the map $(x, u) \mapsto (x, [u])$ establishes $T^1 M$ as a covering of ΔM with two leaves. Conclude that the map $\overline{\pi}: \Delta M \to M$, where $\overline{\pi}(x, [u]) = x$, is a locally trivial fibration, with typical fiber P^{m-1}.

10. If M is connected, the same happens with TM and ΔM, and also with $T^1 M$ when $m \geq 2$. If M is compact, $T^1 M$ and ΔM are also compact.

11. TM and $T^1 M$ are orientable, independent of the orientability of M. ΔM is orientable if, and only if, m is odd.

12. A *continuous direction field* in the surface M is a correspondence $\delta: x \mapsto \delta(x)$ such that the map $\overline{\delta}: M \to \Delta M$, given by $\overline{\delta}(x) = (x, \delta(x))$, is continuous (a section). The direction field δ is said to be *orientable* when there exists a continuous unit tangent vector field $x \mapsto u(x)$ (that is, a continuous section $x \mapsto (x, u(x))$ of the unit tangent bundle $T^1 M \to M$) such that $\delta(x) = \{u(x), -u(x)\}$ for all $x \in M$. If the surface M is simply connected, every continuous direction field is orientable. Conclude from this that the sphere S^2 (and, more generally, any sphere of even dimension) does not admit a continuous direction field.

13. Give an example of a continuous nonorientable direction field in $\mathbb{R}^2 - \{0\}$.

14. Consider $G = \{\pm 1, \pm i\}$ as a properly discontinuous group of homeomorphisms in S^3. Prove that the orbit space S^3/G is homeomorphic to ΔS^2. Conclude that the fundamental group of ΔS^2 is $\mathbb{Z}_4$.

15. Let $\pi \colon E \to B$ be a locally trivial fibration whose typical fiber is simply connected. Prove that $\pi_{\#} \colon \pi_1(E, x) \to \pi_1(B, y)$, $y = \pi(x)$, is an isomorphism. Conclude that, if $\dim M \geq 3$, T^1M and M have isomorphic fundamental groups. If, moreover, M is simply connected, then $\pi_1(\Delta M) = \mathbb{Z}_2$.

16. The homomorphism of the fundamental groups induced by the fibration $\overline{\pi} \colon \Delta M \to M$ is surjective. If $\dim M \geq 3$, its kernel is $\mathbb{Z}_2$.

17. Consider a continuous direction field δ, tangent to the surface M, and let $\widetilde{M} = \{(x, u) \in T^1M; [u] = \delta(x)\}$. Show that $p \colon \widetilde{M} \to M$, defined by $p(x, u) = u$, is a covering with two leaves and that there exists a continuous unit vector field $\widetilde{u}$ tangent to $\widetilde{M}$ such that $[p'(\widetilde{x}) \cdot \widetilde{u}(\widetilde{x})] = \delta(x)$, where $x = p(\widetilde{x})$.

18. With the notation of Exercise 17, show that the field of directions δ is orientable if, and only if, $\widetilde{M}$ is the union of two disjoint open sets, and the restriction of p to each one of them is a diffeomorphism onto M.

19. Use Exercise 17 from Chapter 7 in order to conclude that a compact surface admits a continuous tangent field of directions if, and only if, it admits a continuous, non-null tangent vector field.

Appendix
Proper Maps

A map $f\colon X \to Y$ between two topological spaces is said to be *closed* when, for every closed subset $F \subset X$, its image $f(F)$ is closed in Y. The following proposition characterizes closed maps. Note that it somehow expresses the continuity of the "inverse map" $y \mapsto f^{-1}(y)$, whose values are sets.

Proposition A.1. *In order that $f\colon X \to Y$ be closed, it is necessary and sufficient that, given arbitrarily $y \in Y$ and an open set $U \supset f^{-1}(y)$ in X, there exists an open set $V \subset Y$ such that $y \in V$ and $f^{-1}(y) \subset f^{-1}(V) \subset U$.*

Proof. (Necessary.) If f is closed then $f(X - U)$ is closed in Y and, since it does not contains y, there exists $V \ni y$ open set in Y, with $V \cap f(X - U) = \varnothing$. This means that $f^{-1}(y) \subset f^{-1}(V) \subset U$.

(Sufficient.) Suppose that the condition is satisfied, and take the closed set $F \subset X$. If $y \notin f(F)$ then $F \cap f^{-1}(y) = \varnothing$; hence the open set $U = X - F$ contains $f^{-1}(y)$. Then there exists an open set $V \ni y$ in Y, with $f^{-1}(V) \subset U$, which means that $V \cap f(F) = \varnothing$. Therefore $f(F)$ is closed in Y. □

A map $f\colon X \to Y$, between two topological spaces, is called *proper* when it is continuous, closed and the inverse image $f^{-1}(y)$ of each point $y \in Y$ is a compact subset of X.

For example, if X is compact and Y is Hausdorff, every continuous map $f\colon X \to Y$ is proper.

The inverse image $f^{-1}(y)$ by the inclusion map $f\colon (0,1) \to \mathbb{R}$ is a compact set, for each $y \in \mathbb{R}$, but f is not closed, therefore it is not a proper map. A constant map $f\colon X \to c \in Y$, defined on a non-compact space

X and taking values in a Hausdorff space Y, is closed but it is not proper because $f^{-1}(c) = X$ is not compact.

Proposition A.2. *Let $f: X \to Y$ be a proper map. If $K \subset Y$ is compact then $f^{-1}(K)$ is also compact.*

Proof. Let $\mathcal{U}$ a covering of $f^{-1}(K)$ by open sets $U \subset X$. For every $y \in K$, we can find a finite subcollection $\{U_1^y, U_2^y, \ldots, U_{n_y}^y\}$ of $\mathcal{U}$ covering the compact set $f^{-1}(y)$ and, by Proposition A.1, an open set $V_y \ni y$ in Y such that

$$f^{-1}(V_y) \subset U_1^y \cup U_2^y \cup \ldots \cup U_{n_y}^y.$$

We can also find points $y_1, \ldots, y_k \in K$ such that $K \subset V_{y_1} \cup \ldots \cup V_{y_k}$. Then

$$f^{-1}(K) \subset \bigcup_{j=1}^{k} \bigcup_{i=1}^{n_{y_j}} U_i^y,$$

which proves the proposition. □

Without imposing some restrictions, the converse of Proposition A.2 is false. But it is valid for most of the reasonable spaces. For example, we have the

Proposition A.3. *Let Y be a space whose topology has the following property: if $A \subset Y$ is such that $A \cap K$ is compact for every compact set $K \subset Y$ then A is closed in Y. Let $f: X \to Y$ a continuous map such that the inverse image $f^{-1}(K)$ of each compact set $K \subset Y$ is compact. Then f is closed. If Y is Hausdorff, then f is proper.*

Proof. Let $F \subset X$ be a closed set. For every compact set $K \subset Y$, $F \cap f^{-1}(K)$ is compact and therefore $f(F \cap f^{-1}(K)) = f(F) \cap K$ is compact. Thus $f(F)$ intersects each compact set $K \subset Y$ in a compact, hence $f(F)$ is closed in Y. □

Corollary A.1. *Let Y be a metrizable, or a Hausdorff locally compact space. A continuous map $f: X \to Y$ is proper if, and only if, for all compact set $K \subset Y$ the inverse image $f^{-1}(K)$ is compact.*

In fact, the topology of a Hausdorff locally compact space or of a metrizable space satisfies the condition of Proposition A.3.

Intuitively, the fact that a map $f: X \to Y$ is proper means that *if x approaches the boundary of the set X then $f(x)$ approaches the boundary* *of* Y. The precise formulation of this statement is given by the proposition below.

Proposition A.4. *Let X, Y be metrizable spaces. A continuous map $f\colon X \to Y$ is proper if, and only if, the image $(f(x_n))$ of every sequence in X without convergent subsequences is a sequence in Y that does not have convergent subsequences.*

Proof. Let f be a proper map. If $(f(x_n))$ has a convergent subsequence, say $f(x'_n) \to y \in Y$, then the set K formed by the elements $f(x'_n)$ and the limit y is compact. All of the elements x'_n belong to the compact set $f^{-1}(K)$ and thus they have a convergent subsequence, which is a subsequence of (x_n).

Conversely, suppose that the condition is satisfied and let $K \subset Y$ be a compact set. Then $f^{-1}(K)$ is closed; if it is not compact, there exists a sequence (x_n) in $f^{-1}(K)$ without convergent subsequences in X. From the condition, $(f(x_n))$ does not have convergent subsequences in Y, and this violates the compactness of K, because $f(x_n) \in K$ for every n. □

Proposition A.5. *Let X, Y be Hausdorff locally compact, but non-compact spaces and denote by $\widehat{X} = X \cup \{\alpha\}$, $\widehat{Y} = Y \cup \{\beta\}$ their Alexandrov compactifications. A continuous map $f\colon X \to Y$ is proper if, and only if, the map $\hat{f}\colon \widehat{X} \to \widehat{Y}$, given by $\hat{f}(x) = f(x)$ if $x \in X$ and $\hat{f}(\alpha) = \beta$, is continuous.*

Proof. We leave the proof as an exercise. □

Proposition A.6. *Let X, Y be two metric spaces without isolated points. If a continuous and locally injective map $f\colon X \to Y$ is closed then the inverse image $f^{-1}(y)$ of each point $y \in Y$ is finite and, as a consequence, f is proper.*

Proof. Suppose, by contradiction, that the inverse image $f^{-1}(y)$ of some point $y \in Y$ is infinite. Since $f^{-1}(y)$ is a discrete subset of the metric space X, we can find for each of its points x, an open set U_x containing x, such that these open sets are pairwise disjoint. Moreover, we may suppose that f is injective in each one of these open sets. Now we select a countable infinite family $U_1, \dots, U_n, \dots$ among the open sets U_x and we set $T_n = f(U_n)$. For each n, y is an accumulation point of T_n. Hence there exist $x_n \in U_n$ and $y_n \in T_n$ such that $0 < d(y_n, y) < 1/n$ and $f(x_n) = y_n$. Thus the set $F = \{x_n; n \in \mathbb{N}\}$ is closed in X but $f(F)$ is not closed in Y, which is a contradiction. □

Bibliography

Bredon, G. 1993. *Geometry and Topology.* New York: Springer-Verlag.

Dieudonné, J. 1960. *Foundations of Modern Analysis.* New York: Academic Press.

doCarmo, M. P. 1976. *Differential Geometry of Curves and Surfaces.* New York: Prentice-Hall.

Eilenberg, S., & Steenrod, N. 1952. *Foundations of Algebraic Topology.* Princeton: Princeton University Press.

Francis, G. K. 1987. *A Topological Picture Book.* New York: Springer-Verlag.

Francis, G. K., & Morin, B. 1979. Arnold's Shapiro eversion of the sphere. *Mathematical Intelligencer,* **2**, 200–203.

Godbillon, C. 1971. *Éléments de Topologie Algébrique.* Paris: Hermann.

Hocking, J. G., & Young, G. S. 1961. *Topology.* New York: Addison-Wesley Publishing Co.

Hopf, H. 1935. Über die Drehung der Tangenten und Sehnen ebener Kurven. *Compositio Mathematica,* **2**, 50–62.

Levy, S. 1995. *Making Waves: A Guide to the Ideas Behind "Outside In".* Natick, MA: A K Peters

Levy, S., Maxwell, D., & Munzner, T. 1995. *Outside In* (Video, 21 min). Natick, MA: A K Peters.

Lima, E. L. 1999. *Curso de Análise, Volume 2.* (5th Edition). Rio de Janeiro: Projeto Euclides, IMPA.

Massey, W. 1986. *Algebraic Topology: An Introduction.* New York: Springer.

Peano, G. 1890. Sur une courbe qui remplie toute une aire plane. *Math. Ann.*, **36**, 57–160.

Phillips, A. 1966. Turning a sphere inside out. *Scientific American*, **2**, 112–120.

Seifert, H., & Threlfall, W. 1980. *A Textbook of Topology.* New York: Academic Press.

Smale, S. 1958. Regular curves on Riemannian manifolds. *Transactions of the American Mathematical Society*, **87**, 492–512.

Smale, S. 1959. A classification of immersions of the two-sphere. *Transactions of the American Mathematical Society*, **90**, 281–290.

Spanier, E. H. 1966. *Algebraic Topology.* New York: Springer.

Whitney, H. 1937. On regular closed curves in the plane. *Compositio Mathematica*, **4**, 276–284.

Index